The Quantum Call to Unity

Author's Note

This is not a story about me.

It is only a glimpse through my perception, one lens among many.

I've chosen anonymity out of reverence.

Because the message matters more than the messenger.

Because the ideas are older than I am.

Because some truths shine brightest when no one claims them.

My work continues lovingly, elsewhere.

And the questions I couldn't answer...I leave to the next author.

Maybe that author is you.

So take what moves you and let it change you.

And when the time is right — *write.*

Tell your story. Speak your questions. Follow the wonder.

Because the most beautiful stories are meant to be continued...

— *Elian Noesis Verion*

Chapter 1: The Quantum Call to Unity

"The same stream of life that runs through my veins night and day runs through the world and dances in rhythmic measures."
- Rabindranath Tagore

Rabindranath Tagore

poet, mystic, and seer of Bengal

whose voice sang of unity between the soul and the cosmos

He was the first non-European to receive the Nobel Prize in Literature

a teacher who saw beauty in both silence and song

He founded a school where knowledge was shared beneath open skies

inviting the world to learn with wonder

His words carried the rhythm of rivers

the stillness of trees

and the boundless spirit of a world imagined as one

Chapter Summary: This opening chapter explores the profound interconnectedness woven through our universe on every scale. We delve into scientific revelations from quantum entanglement linking particles across cosmic distances, to living cells communicating with flashes of light, to the life-giving dance of photosynthesis and the internal clocks that guide all creatures. These phenomena, combined with the elegant patterns we see recurring in nature, illustrate a harmony between scientific truth and spiritual insight. Together they invite us to feel a sense of wonder, recognize the unity underlying all existence, and embrace the responsibility and love that come from realizing we are part of one living tapestry.

Look to the night sky and imagine that every star, every spark of light, and every atom in your body is held by unseen threads. This is not just a metaphor. Deep in the heart of quantum physics, we find a truth stranger than fiction. Two particles, once touched by the same origin, can remain linked across vast distances. Change one, and the other responds, as if they share a memory that space cannot erase.

Scientists have witnessed this strange connection in the stillness of laboratories. What they saw defied ordinary logic. Einstein once called it a spooky kind of action, a mystery unfolding across the silence between things.

To scientists, it is a puzzling discovery. To philosophers and poets, it feels like confirmation of an ancient truth: at a fundamental level, separation is an illusion. In a universe woven together by such threads, we begin to sense that everything might be part of one unified whole.

The idea of unity is not new. Sages and seers throughout history have spoken of an underlying oneness, a fabric of being that connects all things. What is astonishing is that science is now catching glimpses of that fabric.

Entanglement hints that the cosmos behaves less like a collection of isolated objects and more like an indivisible web of relationships. In this web, a change in one strand resonates across the whole. It is as if the universe is a vast cosmic tapestry in which each thread – whether a particle, a person, or a star – affects the tension and pattern of the entire cloth. This realization stirs a deep sense of wonder: could it be that we too are enmeshed in this cosmic web, connected in ways we are only beginning to understand?

Light Within Life

The connections that shape us are not only written across the cosmos. They shimmer quietly in the secret places of our own cells. Within every living being, there is a hidden glow.

Scientists have found that living cells release gentle pulses of light. These faint flashes are called biophotons. Though invisible to our eyes, they can be seen with sensitive instruments in the darkness of a lab. Picture this: trillions of cells within you, each flickering like a candle in a vast and quiet city. It is as if your body carries its own constellation, glowing from within.

Some believe this light carries messages. Cells may use it to guide healing, to grow, to know their place in the whole. Even if we do not fully understand it, the fact that we shine at the smallest scale suggests something profound. We are beings of light, not just in metaphor, but in substance.

Across cultures and centuries, light has been a symbol of spirit, awareness, and life. Now, science shows that this symbol may be something more. Light flows through our anatomy as a real and living thread.

When a seed stirs in soil, it signals with glimmers. When the mind forms thought, neurons spark and send light between them. Life knows how to speak this language. Every organism holds an inner lantern. Together, as an ecosystem, we glow like a shared constellation. The boundary between the light of the stars and the light of our cells begins to fade. All becomes part of one luminous field, quietly alive.

Sunlight and the Leaf

Every morning, a sunbeam travels 93 million miles to kiss the surface of a leaf. In that gentle encounter lies one of the most unifying processes on Earth: photosynthesis. When sunlight meets green, living tissue, a transformation takes place that sustains almost every creature in existence. The leaf, with effortless grace, takes the intangible brilliance of a star and turns it into sugar, into wood, into fruit and into life. Through photosynthesis, plants weave sunlight into the fabric of our world. The oxygen we breathe and the food that nourishes us are gifts of this alchemy.

In a very real way, the warmth of the sun becomes the strength of our bones and the beat of our hearts.

Think of how intimately a tree is linked to the sun: rooted in the earth, yet drinking light from the sky. Through the quiet work of leaves, the energy burning in a distant star is shared with the soil, the squirrels, the birds, and our bodies. Each bite of fruit or grain is a reminder of that linkage like a direct line from the heavens to our cells. The sun and the leaf perform a daily dance that feeds the world. It is a partnership as old as life itself.

And remarkably, even the efficiency of this dance might hint at something deeper. Scientists have found that molecules in leaves can capture sunlight with almost uncanny efficiency, leading some to suspect that plants take advantage of subtle quantum principles to make the most of each photon. Whether or not one grasps the physics, the poetry is clear: life and light move together in harmony. Through photosynthesis, nature teaches us that nothing exists in isolation: the life of a plant, the life of an animal, and the light of the sun are braided.

The Rhythm of Life

As the Earth spins and a new day dawns, an orchestra awakens. Birds begin to sing, flowers slowly unfurl their petals, and humans stir from sleep, all in time to the same celestial rhythm. This is the power of circadian rhythms, the internal clocks ticking inside living beings that are set by the cycle of day and night.

Every creature, from the mightiest tree to the tiniest bacterium, carries an innate sense of time aligned with Earth's rotation. When the sun rises, our blood pressure heightens and hormones surge to ready us for activity; when evening falls, that inner clock releases the chemicals that make us drowsy, guiding us gently into rest. Even plants raise their leaves to catch the morning light and let them droop when dusk comes, following a cycle etched into their very DNA.

These daily rhythms are a testament to unity on a planetary scale. Life does not passively endure the day-night cycle; it embraces it, internalizes it, and moves in concert with it. Imagine for a moment the countless species across the globe including people in cities, wolves in forests, flowers in meadows, and plankton in the sea, with each responding to the same transition from dark to light and back again. There is a marvelous coherence in this shared pulse.

We find that our bodies and minds are tuned like instruments to the sun's schedule, even when we do not consciously notice it. In fact, when researchers kept volunteers in dark caves away from external time cues, their bodies still followed roughly 24-hour cycles, driven by an inner cadence. Life carries the imprint of Earth's spin within it. The circadian rhythm is nature's way of echoing the turning of the cosmos in the physiology of every living thing. It is another thread of connection, binding us not only to each other through shared experiences of waking and sleeping, but also to the planet and the star we orbit.

Patterns of Intelligence in Nature

Look at a seashell spiraling in a perfect golden curve and then gaze at a distant galaxy through a telescope and you will see the same spiral arms unfurling in space. A fern unfurling in the forest presents the geometry of a fractal, the same intricate pattern echoed by the branching of rivers and streams, or the splitting veins in your lungs to carry oxygen to cells. These recurring forms are nature's signatures, appearing across the globe and even across the universe. It is as if the artist behind all life enjoys reusing certain masterful strokes, whether painting with galaxies or with petals.

Such patterns inspire awe because they hint that there is order underlying what might seem like chaos. The sunflower arranges its seeds in two interlocking spirals that follow the Fibonacci sequence, a mathematical pattern, to pack them most efficiently. The human body, too, has symmetry and branching networks that speak the same design language as in trees and lightning bolts.

When we encounter these correspondences, it feels like discovering a secret language shared by all of nature. Is it mere coincidence, or evidence of a deeper coherence?

Many scientists simply see the laws of physics and evolution at play. Many spiritual traditions see the hand of a Creator or the emergence of a universal mind. Perhaps these viewpoints converge: the laws of nature *are* the handiwork of a Creator. What is clear is that from the microscopic to the cosmic, patterns repeat and rhyme, suggesting that everything is connected by participation in the same grand design.

An image juxtaposing a spiral galaxy and a nautilus shell, illustrating the recurring spiral pattern in nature

The more we recognize these echoes and rhythms, the more the boundary between "out there" and "in here" begins to dissolve. We find context for ourselves in the symmetry of a snowflake, the cycles of the moon, and the light within our own eyes. The universe ceases to be a collection of unrelated phenomena and becomes instead a living poem, each part reflecting the whole, each moment a microcosm of the cosmic story.

In this light, unity is not an abstract ideal but a tangible reality that reveals itself everywhere we look. We see that a common thread runs through our hearts, our planet, and our stars. This realization comes with a gentle but profound call. If we truly are that connected; if the same stream of life moves through all; then how should we live? Understanding our unity nudges us toward compassion, for others are not really "other" at all, but extensions of the same existence. It encourages a sense of responsibility, for the welfare of any being or ecosystem becomes intertwined with the welfare of the whole. And it inspires love, the natural response to recognizing beauty and kinship in all things.

Each example in this chapter, from the quantum dance of entangled particles to the spiral dance of galaxies, has been a whisper of the same message: we belong to each other and to something greater that is one and whole. This is the quantum call to unity: a call sounded by the very fabric of reality. As you continue through these words and images, let that call resonate in you. Allow yourself to marvel at the interconnected world we inhabit. Let it remind you that you are a vital note in this grand symphony of life, at once unique and inseparable from the cosmic orchestra. In answering this call, we begin to walk a path of connection, responsibility, and love, ever more aware that we are threads in the same magnificent tapestry.

Chapter 2: Cosmic Perspective – Consciousness Beyond Earth

"To see the earth as it truly is, small and blue and beautiful in that eternal silence where it floats, is to see ourselves as riders on the earth together, brothers on that bright loveliness in the eternal cold—brothers who know now they are truly brothers."

— *Archibald MacLeish*

Archibald MacLeish
poet, playwright, and statesman
whose words helped humanity look back upon itself
from the stillness of space
He reminded us that from a great distance
the divisions we cling to fall away
and what remains is the fragile wonder of shared existence
As a voice for both the arts and the conscience of a nation
he wrote not only with ink but with insight
illuminating our place in the vastness
and calling us to remember
that we ride this bright sphere together

Chapter Summary: In this chapter, we journey beyond Earth to explore how a cosmic perspective can awaken a profound shift in consciousness. We delve into the awe-inspiring *Overview Effect* experienced by astronauts and how their reflections reveal a newfound unity with all life and a deep desire to protect our fragile planet. We examine the convergence of scientific discovery and spiritual insight – from Einstein's reverence for the mysterious universe to modern ideas that consciousness might be woven into the fabric of the cosmos. Throughout, the tone is one of wonder, humility, and hope, inviting readers from all walks of life to feel the intimate connection between our minds, our world, and the vast universe beyond.

The Overview Effect: Awe and Unity from Space

When humans first left Earth and looked back from the void, they experienced something extraordinary. Astronauts describe a *cognitive shift*. There is a moment of clarity and transformation now known as the Overview Effect, wherein seeing our whole planet from space triggers *"a state of awe with self-transcendent qualities"*. In the black silence of space, the Earth appears as a luminous blue gem, tiny yet breathtaking, floating in infinite darkness. This visceral view has repeatedly overwhelmed spacefarers with intense emotion and a sudden recognition of oneness: all of humanity and life clinging to the same beautiful world, indivisible by the borders and boundaries that once seemed so important. In that instant, differences of nationality, politics, and creed fade against the backdrop of a *cosmic unity* that encompasses every living being..

One of the first people to experience this shift was Apollo 8 astronaut Bill Anders, who, upon witnessing *Earthrise*, was so awestruck that he impulsively snapped a photo that would become legendary. Anders later reflected, *"We came all this way to explore the Moon, and the most important thing is that we discovered the Earth."* His words capture the essence of the Overview Effect: the real voyage of discovery was not in finding new worlds, but in *seeing our own world anew*.

Astronaut Jim Lovell, orbiting the Moon on Apollo 8, described *"the vast loneliness"* of space and how it made him realize the true value of what we have on Earth. He saw *"a grand oasis"* teeming with life amid an inhospitable void. Similarly, Apollo 11's Michael Collins was struck by Earth's vulnerability: *"It projected an air of fragility... I had a feeling it's tiny, it's shiny, it's beautiful, it's home, and it's fragile."*

Apollo 14 astronaut Edgar Mitchell spoke of an *"instant global consciousness"* that came over him. He had a profound sense of connection with all people and life forms, accompanied by an intense dissatisfaction with the state of the world and a compulsion to do something about it. He described his view of Earth from the Moon as a *"powerful shift in consciousness, emphasizing the interconnectedness of all life"*. This epiphany led him to advocate for a new perspective on human society, one in which we recognize our shared destiny and the need to live in harmony with our planet and each other.

Our conflicts and egos seem petty from a height of 200,000 miles. The famous "Blue Marble" photograph also gave all people a taste of this overview effect, evoking widespread awe and becoming a symbol of the environmental movement. For the first time in history, humanity could see itself from the outside as a single globe without national borders. This new vision nurtured a growing sense that we *all* share one world. Earth Day, global ecological awareness, and even the rise of international cooperation carry echoes of that cosmic perspective. We must come together to protect this little world, the only home we have ever known.

Awe and Humility: When Science Meets Spirit

Confronting the vastness of space and the unity of Earthly life naturally leads to questions beyond science and questions of meaning, purpose, and consciousness. The Overview Effect often leaves us groping for spiritual or philosophical language to express what we feel.

Albert Einstein famously argued that this kind of cosmic awe is central to the human spirit. Though a physicist by training, Einstein spoke often of a *"cosmic religious feeling"* not tied to any organized religion, but a spiritual sense of admiration for the order and mystery of the cosmos. He believed that the pursuit of scientific truth naturally fosters humility and reverence. *"The human mind, no matter how highly trained, cannot grasp the universe,"* he wrote, comparing our intellect to a child wandering in a vast library of which we can only dimly suspect the order. Faced with the *"marvelous structure of reality"* and *"the mystery of the eternity of life,"* Einstein felt only gratitude and awe, coupled with a *"single-hearted endeavor to comprehend a portion... of the reason that manifests itself in nature."* In other words, the more he learned about the universe, the more he sensed something profoundly meaningful at work and something that inspired wonder.

Einstein warned that *"he who can no longer pause to wonder and stand rapt in awe is as good as dead."* For him, maintaining a childlike curiosity and astonishment at the cosmos was essential not just to scientific discovery, but to being fully human. This attitude bridges the often-perceived divide between science and spirituality. It suggests that exploring the stars or the quantum realm can be a deeply spiritual act, cultivating the same sense of reverence that one might find in a cathedral or a forest. Indeed, astronauts returning from orbit sometimes sound like poets or prophets with their scientific mindset expanded to include a newly holistic outlook. They return not only with data and photographs, but with *"eyes turned inward"* often questioning how we live, what matters most, and how our consciousness itself is transformed by touching the wider universe.

Crucially, the cosmic perspective also highlights the limitations of a purely materialist worldview. Modern science, especially since the Enlightenment, has often approached nature as a collection of separate parts to analyze. This mindset called *materialism* holds that only physical matter and forces are fundamentally real. This view has brought great technological progress, yet it can leave little room for the inherent meaning in the cosmos. But when an astronaut suddenly *feels* the living wholeness of Earth, or when a quantum physicist finds that the act of observation seems to influence whether a particle behaves like a wave or a particle, then cracks appear in the facade of pure materialism. Physicist Niels Bohr once remarked that those not shocked by quantum mechanics haven't understood it. In quantum experiments, the observer cannot be wholly separated from the observed. In other words, how we choose to measure something, actually changes what we see. This astonishing fact hints that mind and matter may be entangled at the deepest level of reality therefore challenging the old idea that consciousness is just an accidental byproduct of brain chemistry.

Recognizing these mysteries, some visionary scientists and philosophers have suggested that we need a more expansive understanding of the universe. An understanding that includes life and consciousness as fundamental elements, not mere afterthoughts. The physicist Paul Davies, for instance, has explored the possibility of a "self-explaining universe" with a built-in "life principle," meaning that mind-like qualities could be ingrained in the cosmos itself. Likewise, mathematician and physicist Roger Penrose has argued that new physics may be required to explain consciousness, hinting at Platonic concepts where ideas and perhaps awareness are as primordial as particles. These respected thinkers are not reducing mystery to magic; rather, they acknowledge that traditional materialism's promise to explain everything as purely mechanistic has not been fulfilled. As one commentator put it, when scientists claim that life, consciousness, and even spirituality will eventually be explained in purely material terms, they may be writing a "promissory note" that remains overdue. The cosmic perspective urges us to cash that check differently and to consider that mind and matter are both integral aspects of reality.

Consciousness Woven into the Cosmos

Is it possible that the universe itself is suffused with consciousness? This idea may sound radical, but it is gaining thoughtful attention in contemporary discourse. The view known as panpsychism proposes that consciousness is a fundamental feature of the universe, present even at the level of the smallest particles or fields. In this view, the mind is not an emergent latecomer in an otherwise dead cosmos, but has been there all along, residing in the fabric of the universe. As strange as it may seem, such ideas echo ancient philosophies and spiritual teachings that long held mind and matter to be two sides of the same cosmic coin. Today, some philosophers like David Chalmers (who coined the phrase "the hard problem of consciousness" for the difficulty of explaining subjective experience) have cautiously embraced versions of this view, suggesting that perhaps it's consciousness all the way down. Even a few neuroscientists, grappling with the enigma of how brain activity produces awareness, have mused that if we accept consciousness as a real, irreducible phenomenon, it is a simple step to conclude that the entire cosmos is suffused with sentience.

While most scientists still view consciousness as something that emerges only from particular complex arrangements of matter (like human brains), the panpsychist perspective is invigorating discussions about the place of mind in nature. It invites us to imagine that the spark of awareness we find in ourselves might be a basic aspect of reality at large and that in a very literal sense, we are the universe aware of itself. Such an idea is not easily testable, and it raises many questions (for example, if every electron has a "minuscule mind," how do these combine to form the rich consciousness of a human?). Yet, entertaining this possibility can shift how we relate to the world. If consciousness is woven into the cosmos, then the gulf between "self" and "world" narrows: *we belong here* in a deep and elemental way, as participants in the unfolding story of the universe, not outsiders looking in.

Ultimately, the cosmic perspective is a call to unity. It asks us to embrace a larger identity: not just as citizens of one nation or one planet, but as children of the universe. When we appreciate the rarity and preciousness of our world against the backdrop of countless galaxies, we feel a greater obligation to treat our planet and each other with care. When we recognize that our consciousness may be a cosmic gift, we might approach one another with more compassion, recognizing the spark of the same universal spirit in every person we meet. In this sense, science and philosophy can come full circle.

As we carry that awareness forward, we find ourselves at the threshold of a new way of thinking and being. The Quantum Call to

Unity is an invitation to expand our awareness beyond the old limits and to feel the quantum interconnectedness that threads the universe together, and to answer with acts of empathy and stewardship. In the silence of space, our astronauts found communion with each other, with home, and with the mystery of existence. It falls to us to cultivate this cosmic consciousness in everyday life.

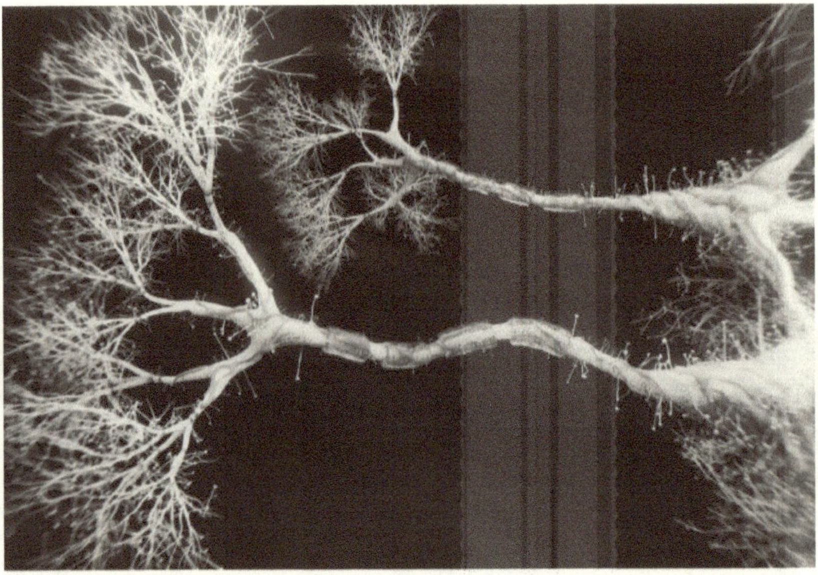

This digitally enhanced rendering of neurons, with its axons and branching dendrites, serve as a conduit of thought and memory.

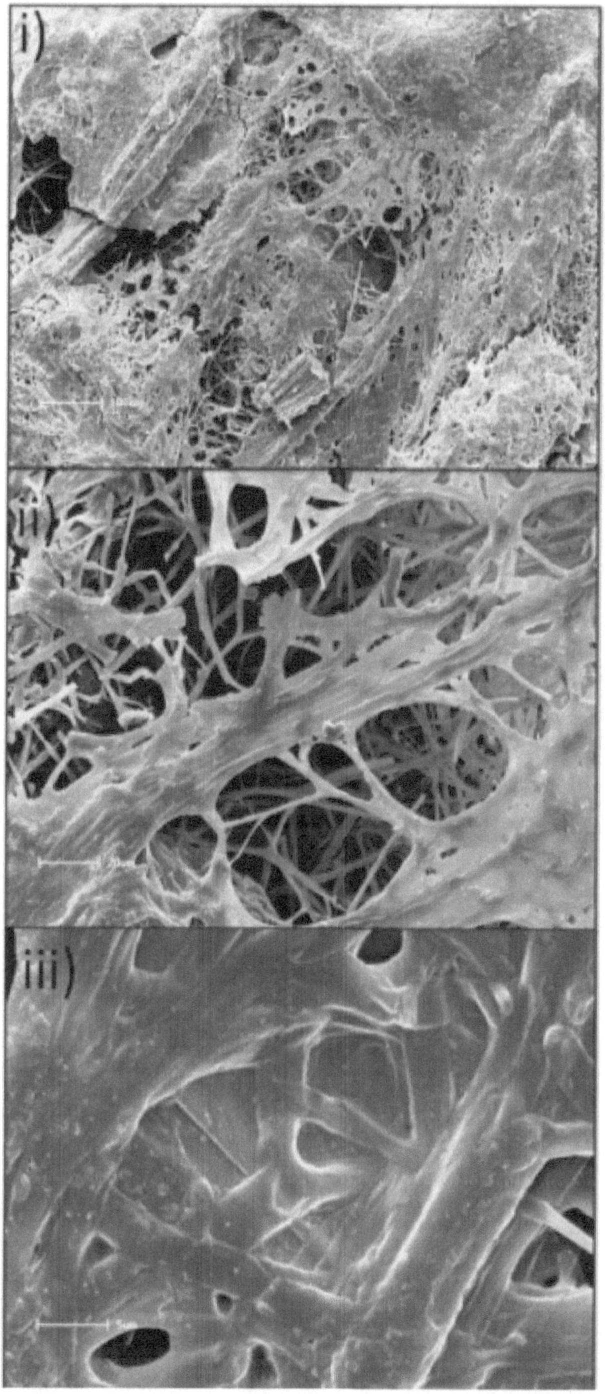
i)
ii)
iii)

Goetten de Lima, Gabriel & Schoenherr, Zaira & Magalhães, Washington & Tavares, Lorena & Helm, Cristiane. (2020). Enzymatic activities and analysis of a mycelium-based composite formation using peach palm (Bactris gasipaes) residues on Lentinula edodes. Bioresources and Bioprocessing. 7. 10.1186/s40643-020-00346-2.

Chapter 3: Quantum Biology

"You are not a drop in the ocean. You are the entire ocean in a drop."

— Jalāl ad-Dīn Rūmī

Rūmī
sage of the soul and poet of the eternal
whose words dissolve the walls between self and source
Born in thirteenth-century Persia
he spoke in the language of the heart
where silence carries more than speech
and a single breath can contain the cosmos
His teachings remind us that the vastness we seek
is already within us
and that every form is a mirror
reflecting the whole from which it flows
In the mystery of quantum biology
his voice returns as a whisper
calling us to see the invisible threads
that bind all things in unity

Chapter Summary: This chapter explores the emerging field of quantum biology and how life's processes interweave with quantum phenomena. We delve into the quantum coherence that aids plants in harvesting sunlight, the quantum entanglement that may guide a bird's migration, and the ultraweak biophoton emissions by which cells might communicate. We consider the possibility of quantum processes in cognition and marvel at fractal patterns repeating across scales in living systems. Throughout, we blend cutting-edge scientific discoveries with philosophical reflection, highlighting the profound interconnectedness and sacred intelligence manifested in nature's design.

Quantum Coherence in Photosynthesis – The Leaf's Hidden Symphony

In the quiet green of a leaf, a subtle quantum dance unfolds. When a photon of sunlight strikes a chlorophyll molecule, nature does something astonishing: the energy doesn't take just a single path to the reaction center of the plant but explores many paths at once in a wave-like superposition. Researchers discovered evidence that photosynthetic organisms exploit quantum coherence – simultaneous multiple pathways – to transfer energy with remarkable efficiency. In essence, the excitation energy "waltzes" through the light-harvesting complex, sampling different routes near-simultaneously, as if the leaf finds the quickest way by feeling all ways at once. Scientists observed long-lived quantum oscillations in photosynthetic complexes at ambient temperatures, indicating that the energy transfer in bacteria and plants truly involves this quantum choreography.

Over the past decade, a vigorous debate has ensued on whether this coherence actively enhances photosynthetic efficiency. A study in 2021 directly probed this question and found that while coherent vibrations occur, they likely do not substantially boost the overall efficiency of energy transfer. In other words, the plant may not *need* quantum tricks to survive yet the tricks are happening anyway. The mere fact that quantum coherence persists in the warm, wet environment of a cell for even a few hundred femtoseconds is astounding. Pigment molecules remained in tune with each other in a unified quantum state just long enough to make a difference.

From a philosophical view, this hidden symphony in the leaf speaks of a deep harmony between life and the physical universe. The sunlight that feeds a plant is literally a vibration of the cosmos, and the plant's molecules answer in the same language which can be defined as a quantum language of waves and possibilities. It is as if in every leaf, there lies a fragment of cosmic wisdom, echoing the unity of existence. The coherent dance of molecules underpins the very visible growth of a tree, blurring the line between biology and physics. In that blur, one can sense a poetic truth: life is not separate from the laws of nature but is woven intimately into them, a song within the cosmic song.

Quantum Entanglement in Bird Navigation – The Avian Compass of Light

On a starry night, a robin embarks on a journey guided by an invisible map. For centuries, humans have wondered how migratory birds know their way across ocean and continent. Beyond sun and star cues, birds possess a mysterious magnetic compass sense. Today, scientists believe that a bird flying in the dark might literally be sensing Earth's magnetic field through quantum entanglement happening inside its eyes. In the retina of birds like the European robin, special proteins called cryptochromes undergo a chemical reaction under blue light that yields a pair of electrons in a quantum-entangled state. These two electrons, though separated by different parts of the molecule, remain entangled, their quantum spins delicately correlated. As the entangled pair "waltzes" between singlet and triplet states, an external magnetic field (even as weak as Earth's) tilts the odds of each state. In essence, the bird's chemical compass uses a quantum process: the entangled electrons' response to the Earth's field alters the chemistry in the bird's eye, giving it a visual or neurological signal of direction.

The evidence for this almost unbelievable mechanism has been growing. Experiments have shown, for example, that disrupting birds with specific radio frequency noise, essentially jamming the quantum dance, can confuse their navigation. European robins lose their ability to orient when exposed to weak, fluctuating radio waves, whereas shielding them from such electromagnetic "noise" restores their compass. This is compelling evidence that a radical pair quantum mechanism underpins the bird's magnetic sense. And in 2021, researchers for the first time observed a telltale quantum behavior (a "non-classical" response to magnetic fields) in actual cryptochrome proteins from night-migratory birds. These findings suggest that the bird's compass is indeed rooted in quantum entanglement, not just a classical iron filing in the beak or brain. Astonishingly, similar proteins in humans have shown the basic ability to sense magnetic fields in laboratory conditions, hinting that the quantum magnetosense might be more common in nature than we ever imagined.

Consider what this means in a broader sense. A tiny robin, with a brain no larger than a walnut, is aligning itself with planetary and atomic forces simultaneously. Life is literally intertwined with the planet: a bird's internal molecules feeling the whisper of Earth's field lines through quantum linkages. Philosophers and poets have long imagined a oneness between living creatures and the Earth; here we have a scientific tableau of that unity. The bird and the Earth are connected by invisible threads of entanglement, a relationship so subtle that the slightest disturbance can sever the compass. There is wonder and humility in this realization. A migratory bird, in its own humble way, partakes in quantum physics to fulfill its instinctual quest. This unity of bird and planet through quantum phenomena invites us to see nature itself as a web of connections, alive with a kind of quiet intelligence that guides even a bird in flight.

An illustration of a night-migrating bird with a faint overlay of Earth's magnetic field lines and entangled electron pairs in its eyes, symbolizing the quantum compass guiding its way.

Biophoton Emission – Cells Communicating in the Language of Light

Not all communication in nature is done with sound or chemicals; some of it is done with light. All around and within us, living cells emit a constant, ultra-dim glow which are packets of light called biophotons. This light is so faint that it's near the threshold of detection, and for a long time, science paid little heed. Yet, decades of research have confirmed that virtually every living cell emits photons as it goes about its metabolic processes. Today, a radical idea is taking hold: cells might use this light to communicate with one another, functioning as a parallel signaling system alongside chemicals and nerves.

In our brains, neurons flash with electrical impulses, but astonishingly, they may also whisper to each other in light. Biophysicists have found that neurons can produce photons during activity and have hypothesized that these photons could carry information through tissue. Imagine a neuron deep in the darkness of the cortex, firing an electrical signal down its axon while simultaneously releasing tiny bursts of photons. Neighboring cells might absorb those photons via light sensitive molecules, tuning their behavior in response. Experiments show that if you separate cells in different containers but allow only light (and no chemical) to pass between them, then stimulating one cell can induce a reaction in the other; as if light itself carried the message. In one striking scenario, when neurons are stressed or damaged, they increase their photon emissions; nearby neurons seem to react to this subtle glow by activating their own defense and repair mechanisms. This so-called "bystander effect" implies that cells are eavesdropping on each other's light signals to warn of danger or injury. Even microtubules (the same structures implicated in quantum cognition) have been suggested to act as waveguides for conducting light within neurons, possibly channeling biophotonic signals efficiently across the cell.

How extraordinary to think that, within the body, there is a silent light based internet linking cell to cell. Each cell is like a tiny star, emitting photons that may be received by other cells as information. In this view, our bodies shimmer with an invisible luminescence: a dynamic field of light that could be coordinating growth, healing, and perhaps other biological rhythms. This concept resonates deeply with spiritual imagery since many traditions speak of an "inner light" or aura in living beings. Here we find a scientific parallel: a literal light inside, connecting the community of cells that comprises a living organism. It underscores the interconnectedness at the cellular level, an interconnectedness so complete that cells might be in constant conversation, not only through biochemistry but through biophysics. It is as if, beyond the genome and the proteome, there is a *"photome"* – a layer of information in the form of light, linking and harmonizing the whole. Such a discovery expands our sense of wonder on the body, revealing another facet of the sacred intelligence of nature.

Fractals and Self-Similarity – Nature's Repeating Patterns

Wander through a forest and you'll notice a curious geometry in the foliage: the shape of the whole appears echoed in the parts. Fractals are the name we give to such self-repeating patterns. The fern's frond is composed of mini-fronds (pinnules) that look like tiny copies of the whole fern leaf. A Romanesco cauliflower is a spiral of spirals with each floret a miniature cauliflower head, arranged in a logarithmic spiral that itself is made of even smaller spirals. Even the branching of a tree to trunk to limbs to twigs, follows a fractal-like algorithm, as does the human circulatory system branching from arteries to capillaries. In nature, fractals arise because they are efficient. This geometry allows maximum surface area in minimum space (think of the fractal branching in lungs creating a tennis court size surface inside your chest for gas exchange). Fractals also allow similar control or function at different scales. For instance, the electrical signals in your heart form fractal patterns in time; a healthy heartbeat has a finely irregular, self-similar variability, whereas a too-regular heartbeat can be a sign of stress or illness. It's as if life knows that to endure and adapt, it must echo its forms across scales, building robustness through repetition with variation.

Discoveries show that nature's affinity for fractals goes down to the molecular level. In an astonishing 2024 study, scientists reported the first ever observation of a regular fractal pattern in a protein molecule. They found that an enzyme in a certain cyanobacterium can spontaneously assemble into the shape of a Sierpiński triangle, a famous fractal pattern consisting of a triangle subdivided recursively into smaller triangles. Under the electron microscope, this protein lattice displayed repeating triangular voids and motifs, the same pattern at multiple scales. This was completely unexpected. While fractal like forms are common in large scale biology, a perfect fractal at the nanoscale had never been seen. Is it classified as an accident? The fractal enzyme happens to have subunits and bonding angles that tile space in a self-way. Yet, the fact it exists at all is breathtaking. It hints that the self-organizing principles of matter can mirror the patterns we see in galaxies and coastlines, right down to proteins. Fractals might emerge whenever systems iterate simple rules with feedback. In biology, this often comes from growth processes such as tiny fern bud growing out to repeat the instructions "make a little copy of me" over and over. DNA itself encodes recursive processes that can generate fractals. The fractal enzyme shows that even without a direct biological function for self-similarity, nature's creative play can produce deep mathematical beauty.

The prevalence of fractals in life invites reflection on a classic idea: *"As above, so below."* The microcosm reflects the macrocosm. Patterns repeat, with variations, from the very small to the very large. This speaks to a profound unity in nature's design. Fractals show how a simple rule can generate inexhaustible complexity and beauty, much as a simple musical theme can lead to endless variations. Is it any wonder that fractal patterns often mesmerize us and even remind us of sacred art? The mandalas of certain spiritual traditions, the intricate patterns of Islamic art, the recursive motifs in Gothic cathedrals all mirror fractals, intentionally or not, perhaps because we intuit the fractal nature of reality. When we recognize fractals in clouds, mountain ranges, blood vessels, or neurons, we get a sense that Nature is self-similar at all levels.This realization reinforces the theme of interconnection: the part contains something of the whole. Each human being, each living creature, might likewise contain a reflection of the cosmos. In fractal geometry, we find a bridge between the quantitative rigor of math and the qualitative experience of awe. Nature's repeating patterns are like a form of visual music, and our ability to perceive them fills us with a sense of belonging. It's as if we too are notes in that endless music of the universe, each note a variation on a theme.

Fractal Self-Similarity Collage

Composite Image (Fractal Patterns in Nature and Math):

Romanesco Broccoli

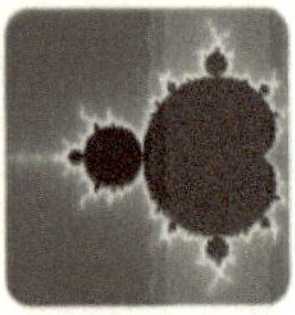

Mandelbrot Set

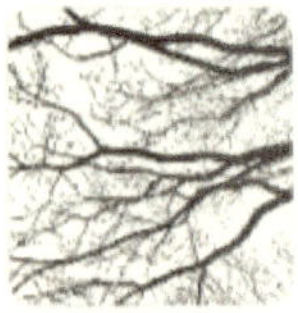

Tree Branches

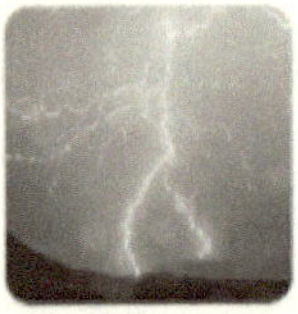

Lightning Bolt

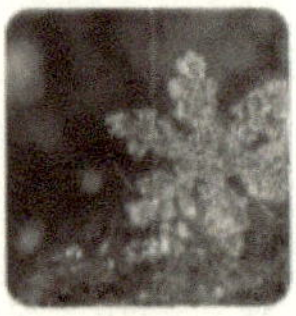

Snowflake

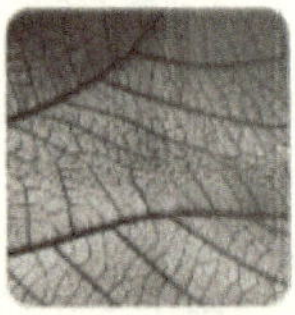

Leaf Veins

River Delta

Nautilus Shell

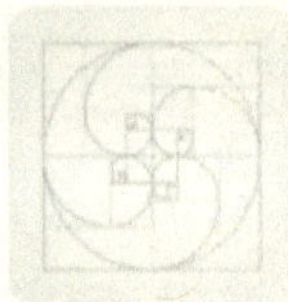

Golden Spiral

Fractal Elements in the Collage:

Romanesco Broccoli

Within this plan, each spiral floret is composed of smaller replicas of the whole, repeating self-similar patterns at diminishing scales. It grows in a logarithmic spiral arrangement, illustrating the Fibonacci sequence in nature. The entire curd displays the same geometry as its parts, a hallmark of fractal design.

Mandelbrot Set

This is a classic visualization of a mathematical fractal. The Mandelbrot set's boundary exhibits infinite self-similarity. No matter how much one zooms in, new mini-brot patterns appear that repeat the whole shape. It represents a pure fractal generated by iterative equations, in contrast to the imperfect but recurring fractals found in nature.

Tree Branches

The limbs of a tree split into smaller branches and twigs in repeating patterns. Each large branch gives rise to smaller branches in a way that resembles the full tree. This branching form is a common fractal in nature. The shape of one twig often mirrors the shape of the entire organism.

Lightning Bolt

Lightning displays branching structures that resemble trees and roots. As electricity travels through the atmosphere, it splits into smaller filaments. The result is a repeated branching pattern created by natural electrical breakdown. This is another clear example of fractal behavior.

Snowflake

Snowflakes grow through repetition. Each crystal develops arms that split into sub-branches while maintaining sixfold symmetry. These smaller features resemble the larger shape, creating a form of fractal self-similarity. Every snowflake is unique but follows the same rule of recursive growth.

Leaf Veins

The network of veins in a leaf forms a repeating structure. A central vein gives rise to secondary and tertiary veins in a branching hierarchy. These patterns are efficient for nutrient delivery.

River Delta

Seen from above, river deltas reveal natural branching systems. Large rivers divide into smaller distributaries, which divide again into channels and streams. This process creates a self-similar structure across different scales. River networks are often studied as geological fractals.

Nautilus Shell

The nautilus shell contains a series of chambers that grow outward in a spiral. Each chamber maintains the same shape while increasing in size by a constant ratio. This is an example of logarithmic growth and approximate self-similarity. Though not a perfect fractal, it reveals nature's tendency toward proportionate and patterned expansion.

Golden Spiral

This spiral appears throughout nature and mathematics. It follows the golden ratio and resembles patterns seen in shells, hurricanes, and galaxies. In this collage, the golden spiral helps link the Romanesco and nautilus shell visually and mathematically. It represents the beauty of repeated growth guided by proportion.

Interconnection, Evolution, and Sacred Intelligence: Weaving it All Together

Threading through all these phenomena is a unifying message: life is deeply interconnected at every level. The quantum coherence in a leaf connects the plant to the sun's quantum light; the bird's entangled compass connects its mind to the Earth; the biophoton signals connect cell to cell in a body; the quantum vibrations in neurons hint at a connection between consciousness and the fabric of reality; the fractal patterns connect each part of an organism to the whole, and the organism to universal patterns. These are not isolated curiosities, but facets of a larger truth: that life emerges from and expresses the fundamental workings of the universe. It's as if nature, in trial and error, remembers what works (a kind of learning), and what works often resonates with deeper laws, whether it be a quantum effect that improves efficiency or a fractal form that optimizes growth.

Standing back, we can see life on Earth as a coherent, intelligent expression of the universe. The quantum and fractal phenomena in biology hint that the boundary between living matter and the rest of the cosmos is thin. Life takes advantage of whatever the universe offers whether in waves, particles, fields, or in geometry, and in doing so, blurs the distinction between the laws of physics and even the behavior of organisms. Perhaps this is the ultimate call to unity that quantum biology sings: the recognition that we are not apart from nature, but a development within it, arising from the same dances of light and matter that shape all things.

Such a perspective has both scientific and spiritual richness. Scientifically, it challenges us to rethink reductionism; we find that to truly understand life, we must sometimes follow it down the rabbit hole into quantum realms or up into abstract mathematical patterns. It encourages a holistic view, where the interactions and connections are as crucial as the individual parts. Spiritually or philosophically, it rekindles a sense of wonder and reverence. The phenomena we've explored in this chapter can feel like modern myths or stories in which the protagonists are photons, electrons, cells, and patterns, all conspiring to allow life to blossom and become aware. These stories invite us to imagine that the universe is alive in us, and we are alive in the universe, inseparable and mutually listening and illuminating.

In the end, quantum biology, life's interconnection, teaches us a beautiful lesson that at every scale, from the tiniest quanta to the broad sweep of ecosystems, there is a play of relationship, a communion of parts. Nothing exists in isolation. The leaf needs the sun, the bird needs the magnetic earth, the cell needs the light of its neighbors, the mind needs the quantum flicker of its microtubules, the organism mirrors the cosmos. This intricate unity is not only the domain of esoteric science; it's also a source of deep meaning. It suggests a cosmos that strives toward life and awareness, that perhaps cherishes connectivity and coherence the way we cherish love and understanding.

As we close this chapter, we stand in awe at the edge of knowledge, peering into a reality where biology and physics merge into a feeling of *life is music and the universe is the instrument*. The melodies of photosynthesis, navigation, communication, cognition, and growth are all variations on a grand theme of interdependence. The more we learn about these quantum and fractal underpinnings, the more we might feel what mystics have long told us that there is an underlying unity to all existence. As we gather glimpses of that unity in measurable, empirical ways, it deepens the mystery and majesty of life. Truly, the quantum call to unity is a call to recognize ourselves as part of an unbroken whole, living participants in a cosmic dance that is at once scientific and sacred.

Chapter 4: Systems Thinking – Patterns, Feedback, and the Wisdom of Wholes

A spider's web beaded with morning dew. Each droplet suspended in delicate balance, a quiet geometry shimmering in the light, capturing the stillness between motion.

"Humankind has not woven the web of life. We are but one thread within it. Whatever we do to the web, we do to ourselves. All things are bound together. All things connect."

— **Chief Seattle**

Chief Seattle
leader, peacemaker, and guardian of the land
who spoke for the Earth long before it was called a resource
A voice of wisdom from the Pacific Northwest
he reminded us that the land does not belong to us
we belong to it

To him, every stream carried memory
every tree stood as ancestor
every creature was kin
His words are not fragments of the past
they are instructions for the future
He offers the ancient truth
that all things are bound together
and to care for the web is to remember who we are

◎ ∞ ◎

You are not just in the web. You are the web.

Chapter Summary: This chapter explores the hidden language of numbers as a doorway into systems thinking and the profound interconnectedness of life. Across ancient civilizations and sacred texts, we uncover how numbers function not merely as tools for measurement but as symbols of purpose, structure, and harmony. From the Hebrew understanding of function over form to the geometric wisdom of Egypt, Babylon, and India, we reveal how numeric patterns encode the architecture of meaning in creation itself. We examine how these rhythms mirror the feedback loops and emergent behaviors found in ecological systems, cosmic design, and human consciousness. By recognizing the recurring archetypes of numbers such as 3, 7, 12, 40, and 108, we invite a shift from fragmented reasoning to a holistic view. Ultimately, we propose that listening to the language of numbers can cultivate ethical responsibility, reawaken wonder, and guide the transformative redesign of our lives, institutions, and systems.

The Hidden Language of Numbers: Meaning Beyond Measurement

In the ancient world, numbers were not merely instruments of calculation. They were sacred language. In Hebrew, as in many early traditions, numbers carried symbolic, functional, and spiritual weight. To count was not only to measure. It was to name the rhythm of reality itself.

In Hebrew thought, the root meanings of numbers are woven into the very structure of biblical scripture. This is not numerology as superstition. It is numerology as pattern recognition, where divine order is revealed in structure and symbol. The word "count" itself has ancient resonance. In Latin, *computare* means to reckon together, combining *com* meaning with, and *putare* meaning to prune or clear up. In the Hebrew view, to number a thing was to see it rightly, to give it its place within a greater whole.

3 is the number of completeness. It represents harmony between elements, often expressed in patterns of movement or transformation. The third day is the day of emerging. Abraham, Isaac, and Jacob form a sacred triad. So do body, soul, and spirit.

7 is woven into Hebrew thought as a symbol of completeness, spiritual fullness, and sacred rhythm. The creation story unfolds in seven movements, not to mark literal time, but to express the arrival of meaning. Genesis does not describe the making of matter. It reveals the setting of a world in order, a space prepared for presence. The word *bara*, often translated as "create," speaks not of shaping physical things but of assigning purpose. To bring something into being was to give it a role within the whole. Time follows that calling. Function comes first. To rest on the seventh day is not to withdraw from labor, but to declare that all has found its place.

12 represents governance, completeness in community, and structural integrity. Twelve tribes. Twelve disciples. Twelve months. It marks the full expression of divine order among people.

40 symbolizes transformation. A sacred period of testing or preparation. Forty days in the wilderness. Forty years in the desert. Forty days of rain. The number speaks of a season required for growth and purification.

These patterns are not unique to the Hebrew tradition. Cultures across the world, separated by oceans and centuries, discovered similar numeric archetypes.

Egypt

The Egyptians revered the number 10, representing divine order and completeness. Their pyramids were constructed using proportions tied to the constants of pi and phi, seen later in nature and human anatomy. The ten plagues of Egypt mirrored the ten utterances of God in creation. Each number carried covenantal resonance.

Babylon

Babylonians used a base 60 system, which still influences our measurement of time. Sixty seconds, sixty minutes, and the division of the circle into 360 degrees. These units were not arbitrary. Sixty and its multiples reflected cosmic balance. The number 360 represented the full circle of heaven, a celestial perfection.

Maya

The Maya calendar was built on cycles of 13 and 20, forming a 260 day sacred calendar aligned with human gestation. Their use of numbers was not only astronomical but spiritual. They understood time as a living, recursive force.

Greece

The Pythagoreans saw the entire universe as governed by numbers. The tetractys, a triangular figure with ten points arranged in four rows, represented harmony, progression, and dimensionality. Numbers were formed before they were quantities.

India

In the Vedic tradition, 108 is a sacred number symbolizing the universe. One for unity, zero for emptiness or fullness, and eight for infinity or the eternal. Prayer beads, sun salutations, and sacred recitations all reflect this hidden cosmic code.

Across civilizations we find repeating patterns.

3, 4, 7, 10, 12, 40, 60, 108, 360: these are not numbers for calculation alone. They are signs. They reveal an architecture of meaning built into creation itself.

Even in mathematics and nature, we see this resonance. The Fibonacci sequence, the golden ratio, and fractals found in ferns, shells, and galaxies suggest that life unfolds according to invisible rhythms. Numbers are not invented. They are discovered.

This perspective echoes the insight of Galileo Galilei, the father of modern science, who wrote:

"Mathematics is the language in which God has written the universe."

Galileo Galilei saw farther not only through his telescope
but through the clarity of his thought
He listened to the stars as if they were speaking
not in words
but in number
in pattern
in motion
To him, mathematics was not invention
but revelation
a way to read the mind of creation itself
He challenged the boundaries of his time
not to conquer truth
but to align with it
His courage lit the path for generations
reminding us that the universe is intelligible
not because we impose meaning
but because we awaken to the order already there
written in light
in gravity
in the silent music of structure

$$\infty \bullet \triangle \bullet \bigcirc = 108$$

• ∞ (Infinity): The eternal, the unbounded, the source beyond space and time. A symbol of unity, divinity, and the infinite nature of consciousness.

• △ (Triangle): A symbol of balance and transformation. Often associated with the number three: body, soul, and spirit or creation, preservation, and renewal.

• ○ (Circle): Wholeness, completion, cycles, and the cosmos. Represents eternal return and the unity of all things.

• 108: A sacred number found in many traditions. It reflects cosmic alignment, the link between microcosm and macrocosm, and the harmony of inner and outer worlds.

Infinity, triangle, and circle. The boundless, the balanced, and the whole. When brought into harmony, they reveal the sacred number 108. This is not a mathematical equation but a symbolic one. A sigil of cosmic order. It speaks to the rhythm between the infinite and the embodied, the geometry of existence echoing through mind, body, and universe. Just as 108 links prayer to breath and breath to cosmos, so too does this triad whisper the mystery of all things moving as one.

Venus

Even the heavens move in harmony. Every eight years, the planet Venus traces a near perfect five-petaled rose in the sky when viewed from Earth, a celestial pattern known as the Venus Pentagram. This elegant dance results from Venus's 584-day synodic cycle aligning with Earth's orbit in a 13:8 ratio.

To the ancients, this wasn't mere astronomy. It was poetry written in motion. The five-petal form echoed the golden ratio and the natural geometry seen in flowers, starfish, and pinecones. Venus, the planet named for love and beauty, literally draws a rose across the heavens.

This is not coincidence. This is coherence.

♀ × 8 Years = ⟐

Venus (♀) multiplied by time (8 years) equals a sigil (⟐): a symbol formed through motion, embodying harmony and cosmic design.

This symbolic line captures one of the most elegant and mysterious patterns in our solar system.

♀ represents Venus, the planet associated since antiquity with love, beauty, harmony, and the feminine.

Over the course of 8 years, Venus aligns with Earth five times at nearly the same points in the zodiac. These five Earth-Venus conjunctions form a symmetrical pattern when mapped against the backdrop of the sky.

The result is ⟐, the Pentagram Sigil of Venus:

This is not artistic embellishment, but a geometric reality of celestial motion.

Each point of the star corresponds to one of the five synodic conjunctions of Venus and Earth. The ratio involved, 5 Venus cycles to 8 Earth years, echoes the Fibonacci sequence and golden ratio (5 and 8 are consecutive Fibonacci numbers).

In this way, the equation is both poetic and mathematical:

The path of Venus as observed from Earth over an eight-year period, forming a five-lobed pattern known as the Venus Pentagram.

In this context, Genesis becomes something more. It is not a debate about when, but a declaration of why. It is the inauguration of purpose. Time becomes a vessel. Meaning becomes the foundation.

Creation is not a machine. It is a temple.

Time is not a countdown. It is a rhythm.

The cosmos is not random. It is composed like a song.

The Echo of 6 and 7: Humanity and Harmony

Within the Hebrew structure of thought, numbers also reveal a moral and spiritual tension. None more than 6 and 7.

6 is the number of man. Humanity was created on the sixth day. It represents that which is incomplete, finite, and striving.

66 is the symbol of many men, the multiplication of human effort still marked by limitation.

666 is the number of all humanity estranged from divine function. It represents systemic failure, a world built on the illusion of self-sufficiency. It is a mirror reflecting what happens when we exalt human systems above sacred purpose.

7 is the answer. It is the completion of what 6 lacks.

77 suggests deepened restoration, a re-entry into sacred harmony.

777 represents fullness upon fullness. It is the pattern of divine alignment made complete. It is the seal of sacred rest and cosmic integrity.

These are not numbers of fate. They are numbers of choice.

To live in 6 is to live in separation.

To awaken to 7 is to enter rest, responsibility, and relational harmony.

The ancients did not worship numbers. They listened to them.

Each one a whisper.

Each one a doorway. Each one a reminder that we are part of something patterned, pulsing, and alive.

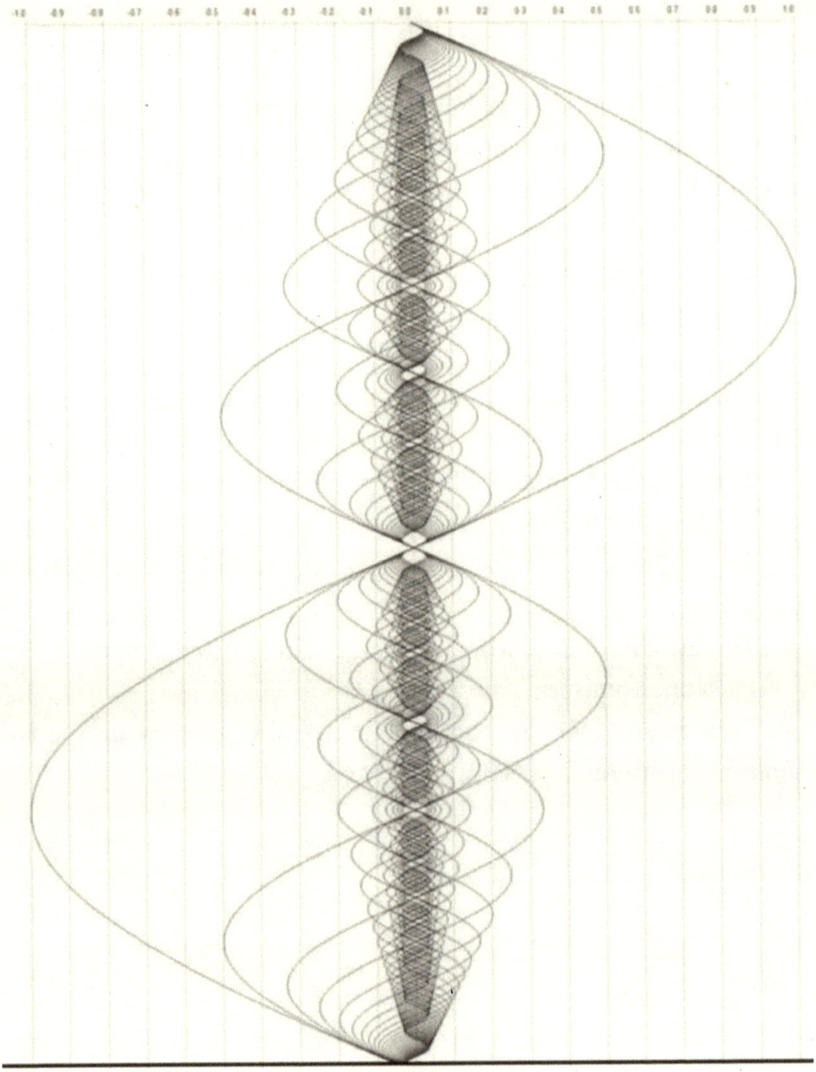

The Music of the Primes

The Music of the Primes

What you see here is not abstract art. It is a glimpse into the hidden architecture of the mathematical universe, a visualization of the

Riemann zeta function, $\zeta(s)$, a mysterious and profound object in number theory. Every dot, curve, and swirl represents a zero of the function, each one plotted in complex space, and together they sketch a portrait of deep symmetry underlying apparent chaos.

This function does more than decorate the landscape of mathematics. It encodes the secret rhythm of prime numbers, those indivisible atoms of arithmetic. The primes seem to appear randomly, scattered without pattern, yet when filtered through the lens of the zeta function, a deeper harmony is revealed. A rhythm too subtle for the eye, yet clear in mathematics.

The image presents a tension between order and disorder. Mathematicians and physicists alike have noted its striking resemblance to quantum spectra, the energy levels of atoms. This is no coincidence. The zeros of the zeta function mirror the behavior of quantum systems, hinting at an unexpected link between the prime numbers and the structure of the physical world.

At the core of this mystery lies one of the most important unsolved questions in mathematics: the Riemann Hypothesis. It suggests that all the nontrivial zeros of the zeta function lie on a single vertical line in the complex plane, the critical line. If true, this would mean that the distribution of prime numbers, though seemingly erratic, follows an exact cosmic order.

This vertical line has been described as a spine of reality, a thread stitching order through numerical infinity. If the Riemann Hypothesis is proven, it will unlock profound insights into everything from cryptography to quantum mechanics. If disproven, it may unravel parts of mathematics we thought were secure.

To contemplate $\zeta(s) = 0$ is to listen for the silence between the notes, the inaudible music shaping the very fabric of number. In this silence, chaos reveals rhythm. In this stillness, the primes pulse like stars in a hidden sky.

Some believe this is the mathematical DNA of the universe, not merely numbers, but structure, resonance, and design woven into the foundation of existence.

$\zeta(s) = 0 \rightarrow$ The Thread Beneath All Number

(Where chaos reveals rhythm and silence counts the primes)

The Web of Interdependence

In a world of intricate relationships, nothing truly stands alone. Systems thinking begins with a simple realization that everything is connected. As naturalist John Muir observed, when we try to isolate any one thing in the universe, we find it "hitched to everything else" around it. In other words, our world is less like a collection of separate pieces and more like a vast, living web of interdependence. A change in one strand of the web can send ripples throughout the whole. This insight invites us to move beyond the old mindset of breaking problems into parts, and instead to see patterns of connection. Indeed, the illusion of separateness is the source of much suffering, as a Buddhist saying reminds us. When we forget how deeply things interrelate, our well-intended actions can backfire in unexpected ways.

Consider a city aiming to save its precious water. Officials encouraged efficient appliances and won a short-term victory: water use per household dropped. Yet soon developers, seeing surplus capacity, built more homes. In the next drought, with no slack left, the city had to drain more from rivers than ever. The system had compensated for our efficiency in a way no one anticipated. Such unintended consequences are common when we don't account for a system's interconnected nature. Systems thinking teaches us to expect feedback and pushback and to act with humility and foresight, knowing our interventions cascade through a mesh of relationships.

Across cultures, metaphors have arisen to capture this truth of interdependence. In the ancient Hindu-Buddhist story of Indra's Net, the universe is envisioned as a vast net of jewels, each reflecting all the others in endless reciprocity. A touch on one jewel is felt across the net, illustrating how every being and event is inextricably interwoven. Modern science echoes this poetic image: ecologists speak of the "web of life," and physicists have even likened Indra's Net to the nonlocal entanglement of quantum theory. We find similar patterns of connection at every scale. The networks of neurons in a brain, the food webs of an ecosystem, the trade links of the global economy are all systems of relationships, patterns that connect beyond individual entities. By observing these patterns, we start to perceive the wisdom of wholes: the idea that there is a collective intelligence or order emerging from the interplay of all parts.

Indra's Net: A shimmering cosmic web of jewels reflects itself infinitely, symbolizing the universe's infinite interdependence with each part containing and affecting the whole.

Feedback Loops: Nature's Cycles of Balance and Amplification

A hallmark of systems thinking is recognizing feedback loops. Feedback loops are the circular chains of cause and effect that regulate or reinforce behavior. Unlike linear thinking ("A causes B, period"), a systemic view sees that B can loop back to influence A. These loops are nature's way of maintaining balance or accelerating change. In our own bodies, for example, countless feedback mechanisms keep us alive: when your blood sugar rises, your pancreas releases insulin to bring it down; when it falls too low, your body triggers hunger to raise it. This kind of balancing feedback (also called negative feedback) pushes against change and holds a system steady, like a thermostat keeping a room at equilibrium.

In contrast, reinforcing feedback (positive feedback) amplifies change. It's a self-strengthening cycle. A classic example is an ice-albedo effect in climate: warming melts polar ice, which exposes dark water that absorbs more heat, causing even more melting. The change feeds itself, growing exponentially until an external limit intervenes. Positive feedback can create virtuous cycles or vicious cycles. Think of trust in a community: acts of kindness build trust, which encourages more kindness, further strengthening the bond of a virtuous circle. Or consider housing market bubbles. As home prices rise, people rush to buy, fearing they will miss out. This surge in demand pushes prices even higher, attracting more buyers and speculators. The rising prices reinforce the perception that housing is a safe bet. But once affordability collapses or interest rates spike, confidence drops, buyers disappear, and the cycle reverses into a crash. Panic selling and foreclosures flood the market, accelerating the decline like a self-reinforcing loop in reverse. In complex systems, these feedback dynamics often determine whether the system stabilizes or spirals into transformation. Scientists describe how certain tipping points occur when reinforcing feedback drives a system across a threshold beyond which changes become self-perpetuating. For instance, beyond a critical temperature, climate feedbacks could lock us into runaway warming. Feedback loops are the heartbeat of system behavior; the pulses that can either sustain balance or unleash rapid shifts.

Crucially, feedback loops often reveal that cause and effect are not neatly separated in time. The effects of an action might boomerang back much later, or at a larger scale. This calls for a longer-term, wider view of responsibility. Systems thinking encourages us to ask: what are the feedback loops here? Will our solution today sow the seeds of tomorrow's problem? By mapping feedback cycles, we become more mindful designers of change, aiming to amplify the balancing loops that promote stability and well-being, and dampen the runaway loops that lead to collapse.

One powerful symbol of feedback is the Ouroboros. Ouroboros is the ancient image of a serpent eating its own tail. The Ouroboros encircles itself in an eternal loop, representing how in a system, outputs are fed back as inputs, creating cycles of cause and effect. In myths, the Ouroboros often signifies wholeness or infinity; in systems, it illustrates how a process can be self-sustaining. Every end becomes a new beginning. This can be the source of resilience, or, if the loop is destructive, the source of persistent problems. We stand within many Ouroboros-like cycles. For instance, consider how consumption feeds production and production feeds consumption in our economy, or how emotions feed actions and actions feed emotions in our personal relationships. Recognizing these loops, we can find points to intervene. Sometimes a simple feedback tweak like a timely reminder or a corrective signal can redirect a cycle from harm to health.

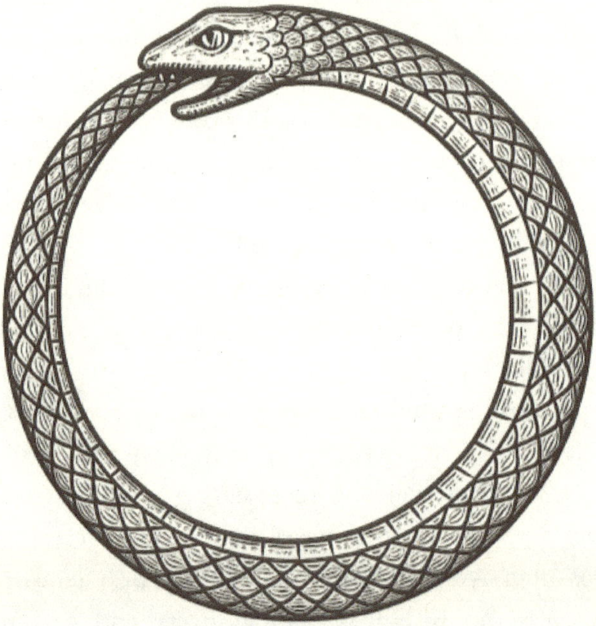

Ouroboros: An ancient serpent devouring its tail, forming a perfect circle. It portrays the self-reflexive cycle of feedback – creation and destruction feeding one another, the system regenerating or consuming itself in endless rhythm.

Emergence: The Whole is Greater Than the Sum

When parts connect and feedback within a whole, something miraculous often occurs: emergence. Emergence is the phenomenon of new properties, patterns, or intelligences arising that none of the parts possess alone. The classic proverb "the whole is greater than the sum of its parts" points to this mystery. A single neuron has no consciousness, yet billions of neurons together give rise to a thinking, feeling mind. A single ant is nearly blind and clueless, yet an ant colony can dynamically allocate foragers, build intricate nests, and respond flexibly to threats. It's as if the colony itself has a mind. In such cases, intelligence and order emerge from the interactions of many simple units. Modern complexity science emphasizes that intelligence (whether in brains, societies, or artificial intelligence) can be understood as an emergent property of complex, dynamic systems. These emergent properties are often unpredictable when examining the parts in isolation; only at the level of the whole do they become apparent.

Life abounds with examples of emergent patterns and intelligence. Watch a flock of starlings swirl at dusk in a murmuration. There are thousands of birds moving as one, weaving fluid shapes in the sky. No single bird orchestrates the dance, yet the flock behaves as if guided by an invisible choreography. It is self-organization in action: simple rules (follow your neighbor, maintain distance, move toward the group's center) are enough to produce a mesmerizing collective performance. Likewise, cells in a developing embryo coordinate to form organs; markets find prices through myriad buyers and sellers; language emerges from a community of speakers innovating and borrowing. Even the patterns of traffic or the growth of cities have emergent qualities resembling ant trails or neural networks. By studying these, scientists and planners glean insight into how order and novelty arise spontaneously from chaos. Emergence is sometimes called the wisdom of crowds, though it can just as easily be the madness of crowds. It depends on the feedback and relationships in play. The key is that new behaviors emerge at the macro level that are not designed top down, but enacted bottom up through countless interactions.

Intriguingly, researchers are finding emergent problem solving in unlikely places. A humble slime mold, for example, can solve complex puzzles without a brain. In one experiment, scientists placed oat flakes (food) to mimic cities around Tokyo, and let the slime mold *Physarum* loose in the center. The slime mold's protoplasm network grew out to connect the food sources and ended up tracing a network nearly identical to Tokyo's rail system. The slime mold had, through simple local decisions, achieved an efficient, robust design that rivaled human engineering. In another study, *Physarum* even solved a maze to find the shortest path, and could compute the optimal route in a traveling salesman problem by dynamically adjusting its tubular network. Such feats astound us because there is no central planner since the slime mold is essentially a single-celled organism spreading out yet its adaptive network behavior shows a rudimentary intelligence emerging from physical feedback processes. Similarly, computer scientists have observed AI agents developing unexpected strategies when interacting in a simulated environment. The behaviors that were not pre-programmed but arose from the dynamics between agents. All these examples underscore a hopeful message: when we connect parts in the right way, collective intelligence can emerge. Solutions to complex problems may be found not by top down control, but by fostering the conditions for bottom up emergence and by allowing diverse agents or elements to self-organize into a coherent, intelligent whole.

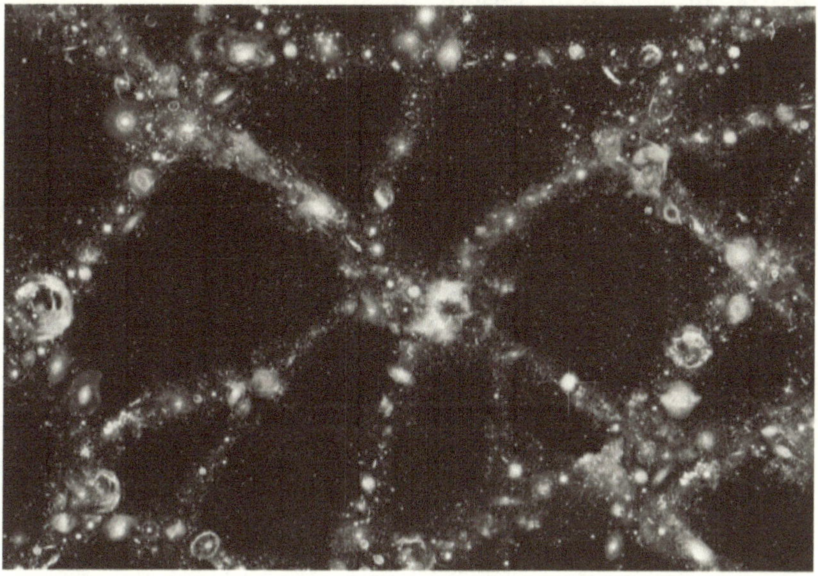

The BOSS (Baryon Oscillation Spectroscopic Survey) map of the universe; a deep space structure visualization. It visualizes the large-scale structure of the cosmos, often called the "cosmic web"

BOSS: The Map of the Unseen

As we step back from the cellular, the neural, and the ecological, a greater picture begins to unfold. Beyond any one part, the universe reveals a structure that is neither random nor rigid, a lattice of meaning woven into the very fabric of space. The Baryon Oscillation Spectroscopic Survey, or BOSS, mapped the distribution of matter on the largest scale ever attempted, tracing not stars or galaxies, but the quiet spaces between them. It is a cartography of void and thread. The cosmos seen not in stars, but in silence between them. What emerged was not chaos, but coherence, a cosmic web of filaments and voids, a fingerprint of gravity stretched across time. Like mycelial networks beneath a forest floor or neurons across a brain, this web reflects something universal, the whole organizing itself without central command. This is emergence made visible. And yet, within this vast structure, smaller systems still move with purpose. Galaxies nest within clusters. Stars nest within galaxies. Planets within orbits. And us, nested within it all.

What we begin to see is a deeper pattern. Not just of size, but of meaning. Not just of scale, but of relationship. A world built from nested wholes. A universe of holons, each part a whole, each whole a part.

Nested Wholes: The Pattern Within Patterns

No part of nature stands alone. Every living system is held within something larger while also made of intricately smaller parts. A cell supports an organ. The organ supports a body. The body belongs to a family, a society, a planet. Each layer is a whole and also a part of something more. Nature, it seems, arranges itself through a rhythm of nested relationships.

Just as earlier chapters explored fractals and quantum coherence as recurring signatures of complexity and harmony, we see a similar cadence here. Systems live within systems, echoing one another in form and function. Each level shapes and is shaped by the others. The architecture of life speaks in quiet recursion, as if the universe were whispering the same idea in different languages.

One powerful framework for understanding these cross-scale interactions is the concept of panarchy from ecology. Panarchy theory tells us that no system can be fully understood by looking at a single scale. A forest, for example, is influenced by processes in individual trees and soil microbes (smaller scale), and by regional climate and nutrient cycles (larger scale). It exists in a hierarchy of scales, each with its own dynamic. Crucially, changes can cascade up or down these levels. Ecologists describe how adaptive cycles (growth, collapse, renewal) occur at different scales and can synchronize or disrupt one another. The forest soil might undergo a renewal cycle annually, while the forest as a whole might have a centuries long growth and disturbance cycle. If a large, slow cycle (say, climate) shifts, it can trigger rapid changes in smaller ones (trees dying back). Conversely, innovative experiments at small scales (new species or practices) can, if conditions are right, revolt upwards to transform the larger system. In a resilient system, these interactions are balanced so that the collapse at one level is contained and becomes an opportunity for renewal, much like a forest fire making way for new growth without destroying the entire forest. Nested hierarchies thus contribute to resilience. They compartmentalize risk and spread adaptation. Our human societies also function this way. For instance, a neighborhood may bounce back from a shock even if the city's economy is struggling, or a global trend might slowly filter down into local culture.

Understanding nested systems encourages us to design with context in mind. A policy that works at a small community level might need adjustment to scale up nationally. Solutions are not one-size-fits-all; they must be tailored to the scale and embedded in larger supportive structures. It also teaches us humility and tells us we are wholes with agency and also parts of greater wholes whose health and purposes we must serve. In many spiritual traditions, this is expressed as seeing oneself as a cell in the body of humanity, or humanity as an organ in the body of the Earth. Each level has its role, and wisdom comes from honoring the whole hierarchy of life.

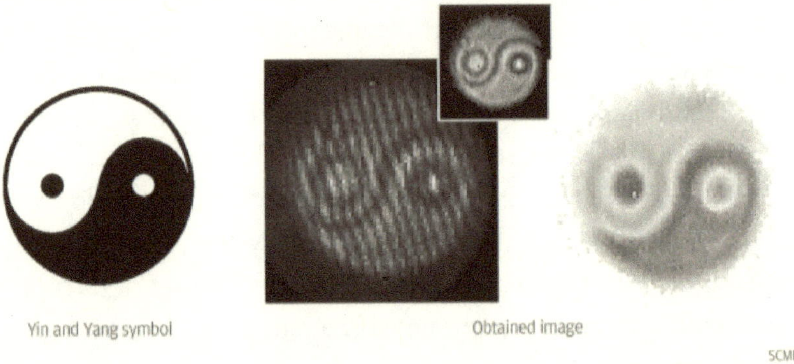

A striking visual parallel between the ancient Yin-Yang symbol and a modern quantum phenomenon. On the left, the Taoist Yin-Yang represents complementary opposites, eternally cycling in unity. On the right, a photograph from a quantum optics experiment reveals the spatial distribution of entangled photon pairs, captured as they exhibit a mirrored, swirling pattern, visually and philosophically echoing the Yin-Yang form.

What appears to be an artistic coincidence is in fact a real, observable outcome of entangled quantum states. The colored image is a two-photon spatial correlation pattern, a direct result of quantum interference, where the probability of detecting one photon depends on the position of its entangled partner.

The photons behave not as independent particles, but as interdependent parts of a single wavefunction, responding to one another across space with unified symmetry. Their distribution doesn't just resemble the Yin-Yang, it embodies it, manifesting an ancient truth at the subatomic level even opposing forces are inseparable and dynamically balanced.

This image becomes a sigil in its own right. It is another bridge between science and spirit, reminding us that nature's most foundational behaviors reflect the same pattern of unity and duality encoded in ancient wisdom traditions.

Co-evolution: Learning and Adapting Together

Life is not static, it evolves. And importantly, it co-evolves. In a system, parts do not evolve in isolation, they adapt in response to one another, often in a dance of mutual influence. Predator and prey species, for instance, continually co-evolve. Faster wolves make for faster deer, which in turn pressures wolves to adapt further. Their evolutionary paths are intertwined. Co-evolution occurs in technology and society as well. Our tools change how we live, and how we live drives the invention of new tools. Consider how smartphones altered communication patterns, which then spurred the development of entire industries (social media, apps) that further changed our daily habits. Humans and our microbiome co-evolve (our diets influence gut bacteria, which in turn influence our health and even behavior, feeding back into cultural food practices). In all these cases, change is a relationship process, not a solitary one.

In human systems, co-evolution often manifests as culture and policy evolving together. For example, as communities embrace renewable energy technologies, their values and lifestyles start to shift (people become more energy conscious, perhaps adopting efficient appliances or shifting to electric cars). Those behavioral changes create political support for greener policies, which then accelerates innovation in clean technology, driving prices down and making adoption even easier. This co-evolutionary loop between society and technology can rapidly transform the energy system as a whole. In fact, some scholars talk about triggering positive tipping points so there are inflection points where a sustainable practice becomes self-reinforcing across society. Such positive tipping dynamics might include the rapid spread of plant based diets, or a cascade of cities divesting from fossil fuels, each change making the next one more likely. Co-evolution is at play in our ethical consciousness, our economic models, and our technologies which all lean forward together. We evolve as a system.

Recognizing co-evolution encourages us to work with the flow of change. Instead of forcing static solutions, we can seed processes that adapt and learn. It also teaches patience since lasting change often requires multiple iterations of mutual adjustment. It's more like a give and take relationship between individuals, society, and environment. In this dance, communication and feedback are key (just as genes communicate via survival success, ideas communicate via social feedback). By staying attuned to how the system responds and evolves, we can become co-creators of evolution, guiding systems toward healthier trajectories.

Resilience, Collapse, and Transformation

All living systems face disturbances. Resilience is the capacity of a system to absorb these shocks and still continue its core functions. It doesn't mean no change; rather, a resilient system can bend without breaking, adapt without losing itself. Ecosystems exemplify resilience: a savanna can endure periodic fires and regrow. A coral reef can bounce back after a bleaching event if given time and reduced stress. A resilient community might endure an economic downturn or disaster but recover through mutual aid and innovation. Systems thinking places great emphasis on resilience because in an interconnected world, disturbances are inevitable but collapse doesn't have to be if we design for resilience.

Resilience often comes from diversity, decentralization, and feedback. A rainforest with many species is less likely to all succumb to one disease. An economy with diverse sectors is less prone to total failure if one sector crashes. Feedback loops detect problems early and correct course, like an immune system nipping an infection. Redundancy provides backups. These traits can be deliberately cultivated. For example, modern electrical grids are being reimagined as distributed networks of many small energy producers (solar panels on homes, community wind farms) rather than a few giant power plants. This makes the grid more resilient to a failure at any single point. Likewise, the internet's packet routing system was designed to reroute information if one node goes down which was inspired by resilient communication.

However, resilience has limits. When a tipping point is crossed, the system's internal feedbacks drive it into a new state often quickly and irreversibly. A lake, for instance, can take nutrient pollution for a while, until a threshold is reached and it suddenly flips from clear to permanently algae choked. The change can become self-sustaining when algae blooms deplete oxygen and kill fish, which further fuels algae with decomposing matter, therefore reinforcing feedback loop locked in a degraded equilibrium. Similarly, climate scientists warn of tipping points like the disintegration of ice sheets or the dieback of the Amazon rainforest, beyond which cascading effects fundamentally alter the climate system. Part of the urgency is that tipping points can come with little warning. A system might appear stable until the moment it lurches into collapse. This is why monitoring and precaution are vital.

Yet, systems thinking also shows that not all tipping points are disastrous. Some can be opportunities for transformation. The concept of creative destruction in economics recognizes that when an old industry collapses it makes way for new industries and innovations. Think of how the rise of digital cameras ended the film camera era, but ushered in a new age of photography. In ecology, a forest fire ravages mature trees but allows sunlight for saplings and returns nutrients to soil, rejuvenating the forest. Socially, a revolution or social tipping point can topple an unjust order and open space for a more just system (though not without risk and turmoil in the transition). The wisdom of wholes perspective reminds us that sometimes a system must break down to break through. Our task is to ensure that when transformation happens, it leans toward life and healing.

In practical terms, fostering resilience means building buffers, feedbacks, and flexibility so that we avoid catastrophic tipping points. For example, curbing greenhouse gas emissions aggressively now is essentially about keeping Earth's systems within stable bounds. On the other hand, pursuing bold change can mean intentionally pushing certain social systems to a positive tipping point, as discussed above, where change accelerates for the good (like a rapid global shift to renewable energy). It's a delicate balance: when to stabilize, and when to catalyze. The art of systems thinking is to know the difference.

A mythical phoenix rises from flames and ash. It embodies the cycle of collapse and renewal, the way a system, when burned to its essence, can be

reborn anew. This sigil reminds us that from every destruction, new life can emerge in a transformed state.

Applying the Wisdom: Systems Thinking in Action

The true power of systems thinking lies in its application which is not simply as a model for understanding, but as a method for healing, building, and transforming. It invites us to see the invisible threads that connect economy, ecology, and human experience, and to design not for control, but for emergence. In recent years, pioneers across sectors have begun turning these insights into living systems such as patterns of policy, technology, and community that work in harmony.

Many of these efforts are inspired by nature. Forests regulate themselves without centralized governance. Wetlands filter toxins, buffer floods, and support biodiversity through layered complexity. Fungal networks exchange nutrients between trees, sustaining life across species. Nature teaches us that resilience is not built through rigidity, but through interdependence, feedback, and adaptation.

Applications of Systems Thinking

• Estonia's e-Governance Platform

Estonia has created a digital ecosystem where healthcare, tax, voting, and business operations are interconnected through its X-Road system. Citizens can file taxes automatically, access health records securely, and manage legal documents without ever waiting in line. This is systems thinking in civic infrastructure with transparency, responsiveness, and human-centered design at a national scale.

• Brazil's Participatory Budgeting (Porto Alegre)

In this globally replicated model, citizens help shape their city's budget through community assemblies. This decentralized, feedback driven structure redistributes power and resources based on local knowledge.

• India's GST & Aadhaar System

India's digital tax system aligns real time business activity with national oversight. Paired with its Aadhaar digital identity network, businesses can reconcile taxes, report income, and access credit more efficiently. This initiative, though imperfect, reflects a shift toward unified, transparent systems that reduce redundancy and fraud.

• Kenya's Regenerative Agriculture Initiatives

Through integrated practices that blend indigenous wisdom, modern soil science, and climate resilience, farmers in Kenya are building circular food systems. These systems increase yield while healing the land, proving that sustainability and prosperity are co-arising goals.

• Canada's CRA API Integration

The Canada Revenue Agency now provides real-time API connectivity to accounting systems, enabling small businesses to automatically sync financial data for payroll, tax filings, and compliance while minimizing errors and freeing time for value creation.

• Taiwan's Digital Democracy

With tools like vTaiwan, citizens engage directly in policymaking through online consensus platforms. Here, artificial intelligence supports human judgment, helping governments remain responsive, adaptable, and in tune with the people they serve.

• IRS Modernization Efforts (U.S.)

The U.S. Internal Revenue Service has begun shifting toward paperless processing and simplified e-filing. The Direct File pilot program and digital correspondence initiatives reflect the early stages of a system that could be designed for accessibility, clarity, and collaboration, rather than confusion and compliance anxiety.

The Call to Reimagine Recordkeeping and Tax Systems

Nowhere is this shift more urgent than in the way we manage and report economic activity. For generations, our recordkeeping and tax systems have been fragmented and riddled with outdated forms, isolated codes, and processes that confuse even the most diligent filers. But what if we approached this differently?

Imagine dynamic, integrated platforms that align small business data with regulatory needs in real time; not as surveillance, but as support. A system designed to reduce mistakes, not to punish them. One that invites clarity rather than confusion. One that sees the small business owner as a trusted contributor to collective prosperity.

These are not distant ideals. Pilot programs already exist. The challenge is not technological, it is moral and structural. It is about will, alignment, and the courage to build something better.

So, what would our institutions look like if they were designed solely for care?

They would be intuitive, transparent, and accessible by empowering users instead of overwhelming them. They would treat honest mistakes as learning opportunities, not liabilities. They would reduce duplicative effort, close gaps between public and private sectors, and use data to serve people.

What if our systems were designed not for compliance alone, but for coherence?

They would unify disconnected processes like licensing, banking, bookkeeping, payroll, and tax into one simple journey. They would communicate with one another. They would listen. A small change in one area would ripple through the whole, eliminating redundancy, revealing insights, and adapting in real time.

What if we designed not just to measure value, but to create it?

Then bookkeeping becomes a form of empowerment. Tax becomes a reflection of shared contribution. Recordkeeping becomes a conversation between enterprise and society. A society that builds trust, confidence, and dignity at every level.

This is the promise of systems thinking: not just merely to understand the world more clearly, but to reshape it more wisely.

And this is the quantum call to unity: to align our structures with our values, our tools with our truth, and our systems with the sacred fabric of life.

In business and technology: Companies are learning that they are not isolated profit machines but participants in broader social and ecological systems. This realization has spurred the rise of circular economies and regenerative design. For instance, instead of the linear "take-make-waste" model, firms are redesigning product life cycles so that waste from one process becomes input for another ecosystem so nothing is wasted. This systemic redesign reduces pollution and creates economic opportunity (recycling industries, biomaterials, etc.). Systems thinking has also entered corporate strategy in the form of ecosystem mapping. Ecosystem mapping is an understanding of the network of stakeholders, resources, and incentives around a product. Tech companies deploying a new app might consider its effects on mental health, community cohesion, or even climate impacts, treating these not as externalities but as integral to the product's design. During the COVID-19 pandemic, many businesses discovered the importance of systemic resilience: supply chain disruptions taught the value of having diverse suppliers and local backups, and some companies adapted by repurposing manufacturing lines or partnering in new ways. In fact, innovation itself can be sparked by a systems view. Seeing the part of the iceberg that's beneath the water." The underlying system structures allow creative solutions that wouldn't be obvious in a silo. One example is the push for "doughnut economics," a framework that guides businesses and cities to operate in a safe zone between social well being and planetary boundaries. It's essentially applied systems thinking for economics, balancing multiple feedback constraints like human needs and ecological limits at once.

In public policy and communities: Systemic approaches are being used to tackle entrenched social issues. Leaders are convening whole system stakeholders to find leverage points. Initiatives in the Netherlands and Scandinavia use "health in all policies" where every policy area (transport, urban planning, finance) is examined for its health impacts on citizens, aiming to create a virtuous cycle of well being. Education reformers, too, are applying systems thinking by redesigning curricula to teach students the interconnectedness of disciplines and real world phenomena. Some schools now have students map systems, like tracing the journey of water from rainfall to tap to ocean, to build systems literacy from a young age. During the pandemic, the countries that fared best often were those who took systemic action by coordinating across healthcare, economic relief, community engagement, and information networks, rather than treating it as a narrow medical issue. This underscores why systems thinking matters today. Our challenges (climate change, pandemics, inequality) are systemic by nature, spanning boundaries and feeding back on themselves. By training ourselves to see those interconnections, we can craft responses that address root causes, not just symptoms.

In design and innovation: A practice called systemic design merges design thinking with systems thinking. Designers using this approach start by mapping the whole context of a problem starting with human needs, then cultural narratives, technology, and into logistics to identify key relationships. Solutions are then prototyped with an eye to how they influence the larger system. For instance, urban designers tackling traffic congestion might integrate public transit planning with housing policy and even virtual work trends, aiming for a systemic reduction in the need for long commutes. Another example comes from agriculture. Regenerative farming practices such as permaculture treat the farm as an ecosystem, integrating crops and animals in complementary ways, building soil (which sequesters carbon), enhancing biodiversity, and yielding nutritious food which is a win-win for multiple parts of the system. These approaches are spreading as evidence grows that holistic solutions are often more resilient and beneficial in the long run. They might require more up front coordination, but they pay off by avoiding the unintended consequences that plague narrow fixes.

Across all these domains, a common thread is emerging: collaboration. Because systems cut across silos, tackling them requires bringing together diverse perspectives from the likes of scientists, business leaders, indigenous elders, community members, and policymakers in a shared learning process. Tools like system mapping, scenario modeling, and organizational constellations are helping people see the system together and find consensus on leverage points for change. This inclusive, participatory aspect gives systems thinking a moral dimension. It values every stakeholder as part of the whole and seeks solutions that consider all voices and impacts, not just powerful or immediate ones.

The Wisdom of Wholes: Unity and Responsibility

As we deepen into the systems view, a profound shift in consciousness can occur. We start to feel the truth of what science and spirituality alike have been telling us. We are not separate from each other or from the larger world, but participants in a flowing, interdependent oneness. This realization is both humbling and empowering. It humbles us because we see that we cannot control everything. Outcomes emerge from the whole, not from our will alone. But it empowers and obliges us because we recognize that everything we do, every choice we make, matters. With awareness of interconnection comes a deepened sense of ethical responsibility.

If "all things are bound together," as the opening quote insists, then to harm any part of the system or another person, a community, an ecosystem is in some way to harm oneself. This mirrors the Golden Rule found in spiritual traditions, now seen through a systemic lens. Why treat your neighbor as yourself? Because at a systems level, they *are* yourself. In essence, they are part of the overlapping wholes that constitute your life. In the environment, this ethic manifests as stewardship. In society, it manifests as striving for justice. An injustice to any group creates turbulence that eventually unsettles the whole society. In our personal lives, it calls for integrity. What we do in private might seem isolated, but it contributes to the field of human consciousness and culture we all share.

Systems thinking thus becomes more than an analytical tool; it becomes a way of being. Some call it systems being or systems feeling. It's essentially an embodied awareness of interdependence that guides our decisions. We move from seeing connections intellectually to sensing them viscerally. This can foster empathy, compassion, and a willingness to work for the common good. We begin to ask: What is the whole telling us? What would benefit the system, not just my piece of it? These questions lead to wise action. For example, a business leader might shift from "How can my company win?" to "How can the ecosystem we are part of thrive (and my company along with it)?" An individual might shift from blame to curiosity. If there's a problem, how have we together created it, and how can we together move forward beyond it?

This mindset resonates with spiritual perspectives. Indigenous wisdom often speaks of "all my relations," extending kinship to all living things. The Earth Charter and other global ethics frameworks explicitly draw on systems thinking, urging that we think of Earth as one community with a common destiny. As we face the planetary crisis of climate change, this outlook could not be more relevant. It is a call to unity: recognition that only by acting in concert, with a sense of shared fate, can humanity navigate the challenges we have wrought. In complexity terms, we may really need is coordinated emergent response at the global level with countless local actions adding up to a civilization wide transformation, a kind of emergent intelligence.

As we answer this quantum call to unity, may we embrace the wisdom of wholes in our minds and hearts. Every pattern is telling a story, every feedback loop is teaching a lesson, every whole is greater than its parts. The more we listen, the more we realize that we, too, are the system. And when we change, the world changes. Let this knowledge guide us to design, to innovate, and to care knowing that in the beautiful, intricate, living system that is our universe, every choice matters, and everything connects.

Chapter 5: Ecology – Stewardship and the Intelligence of Earth

"The major problems in the world are the result of the difference between how nature works and the way people think."

— Gregory Bateson

Anthropologist, Ecologist, and Systems Theorist

Gregory Bateson
a thinker of profound clarity
who saw mind not as a thing but as a pattern
He walked the boundary between science and philosophy
reminding us that to understand the Earth
we must listen not just with logic
but with reverence for relationship
He revealed that nature does not act in straight lines
but in loops of feedback
in webs of meaning
in rhythms that resist control
His work calls us to see the world not as parts
but as living wholeness
and to heal our future
we must first change the way we see

Chapter Summary: This chapter explores ecology as a sacred web of relationships uniting all species, elements, and cycles on Earth. We view the planet as a living, intelligent system that regulates itself through intricate feedback loops and interconnected microbiomes. We examine fragile ecosystems like coral reefs, permafrost, ocean currents, deep forests, and mangroves that hang in the balance, nearing tipping points but also showing signs of resilience and recovery. Grounded in recent science and enriched by spiritual and indigenous wisdom, this chapter calls for an urgent yet hopeful ethic of planetary stewardship, inviting us to belong to Earth's community of life and act with compassion and courage.

Earth as a Living System (Gaia)

Our planet behaves in many ways like a living organism with its own form of intelligence. The Gaia hypothesis, first proposed by James Lovelock and Lynn Margulis, suggests that life interacts with geology, oceans, and atmosphere to maintain habitability. In other words, Earth self-regulates like a coherent being. While once controversial, this view has gained support as we uncover remarkable feedback loops and planetary homeostasis in Earth's past. For example, the composition of our atmosphere (around 21% oxygen) is dynamically stabilized by life. Plants and phytoplankton continually renew the oxygen and absorb carbon dioxide. The planet "breathes" in a seasonal cycle as the vast boreal forests inhale CO2 in summer and release it in winter. It's as if our world has a metabolism.

On a microscopic scale, Earth is knit together by its microbiomes. Just as the human body relies on a gut microbiome for digestion and immunity, the Earth relies on unseen communities of microbes to function. Soil bacteria and fungi recycle nutrients and make the ground fertile; marine microbes generate much of the oxygen we breathe; even the clouds often form around airborne microbes that act as seed coverings for raindrops. In this sense, the living Earth has organs and tissues: forests act as lungs producing oxygen, wetlands and oceans as kidneys filtering water, and fungal networks as neural networks linking roots across an ecosystem.

Modern Earth system science has revealed surprising ways the planet's intelligence manifests. When the Sahara desert blows its dust across the Atlantic, it is not random, it is a lifeline for the Amazon. Each year, millions of tons of mineral rich Saharan dust (laden with phosphorus from an ancient lakebed) travel on wind to fertilize the Amazon rainforest's soils. Likewise, the oceans have conveyor belt currents that distribute heat, nutrients, and salt, preventing extremes of climate in any one place. This global ocean circulation, powered by differences in water density, behaves like the bloodstream of the biosphere. It's humbling to realize that Earth's myriad parts communicate in such ways. The desert cares for the jungle and the ice of the poles speaks to the tropics through sea level and reflection of sunlight. Gaia is not a mystical abstraction but an elegant description of emergent self-regulation arising from countless interactions.

However, like any living body, Earth can be thrown off balance by illness. Our modern civilization's impact is testing Gaia's resilience. Just as a body runs a fever to fight infection, the Earth is heating up under the strain of greenhouse gases. Will the planet's natural intelligence, with its feedback loops and living systems, be enough to restore equilibrium, or will our actions push Earth's ecosystems beyond their capacity to self heal?

Fragile Ecosystems: The Edge of Collapse

Even the strongest webs can fray. Around the world, critical ecosystems that have thrived for millennia are showing signs of stress and fragility under cumulative pressures. Scientific research in recent years highlights several bellwether ecosystems at risk of crossing irreversible thresholds. Below are a few of the most significant fragile systems hanging in the balance:

• Coral Reefs: Often called the "rainforests of the sea," coral reefs cover less than 1% of the ocean floor but support roughly 25% of all marine species. Teeming with colorful fish, mollusks, and intricate corals, a healthy reef is a celebration of biodiversity. Yet rising ocean temperatures and acidification are causing mass bleaching events, where corals expel their symbiotic algae and turn ghostly white. In 2016 and 2017, for example, unprecedented heat devastated huge swaths of Australia's Great Barrier Reef. Without rapid climate action, many reefs could perish within decades. This would be a loss that would echo through the entire ocean food chain.

• Arctic Permafrost: The frozen soils of the far north and high mountains lock away enormous quantities of carbon (about 1,500 gigatons, roughly twice the total currently in Earth's atmosphere). This permafrost has been stable since the last Ice Age, acting as a cold storage vault for ancient plant matter. Now, as the Arctic warms at over three times the global rate, these grounds are thawing. In places, the land is literally collapsing into sinkholes and lakes as ice within the soil melts. Microbes awaken and decompose the once-frozen organic matter, releasing carbon dioxide and methane. This creates a dangerous feedback loop: warming melts permafrost, which releases gases that cause more warming.

• Oceanic Currents: The Atlantic Meridional Overturning Circulation (AMOC), which includes the Gulf Stream, is a crucial global climate regulator, redistributing heat from the tropics toward the poles. In recent decades, evidence suggests the AMOC is weakening, likely due to an influx of freshwater from melting Greenland ice slowing down the "engine" of cold, salty water sinking. Researchers have raised alarms that this current could approach a tipping point where it collapses or dramatically slows, which would wildly disrupt weather patterns across Europe, Africa, and the Americas. While debate continues on the timeline, the trend is concerning. A shutdown of these currents in the geological past triggered abrupt climate shifts. Even a slowdown can lead to more extreme storms, droughts, and sea level rise in certain regions.

• Ancient Forests: Great forests are pillars of planetary stability, pumping moisture into the air, storing carbon, and housing countless species. Today, the Amazon rainforest, the world's largest tropical forest, is nearing a feared tipping point. Ecologists have long estimated that if roughly 20–25% of the Amazon is deforested, the region could dry out enough that the forest can no longer sustain itself. We are perilously close to that threshold; combined with rising heat and worsening droughts, parts of the Amazon are already shifting from lush rainforest to open savanna. In 2024, a study warned that up to half of the Amazon's area could reach a state of rapid collapse by 2050 without urgent action. Similar concerns loom over other old growth ecosystems: the boreal forests of Canada and Siberia are suffering mega fires and insect outbreaks as temperatures climb, and once mighty forests from Borneo to the Congo are fragmented by logging and agriculture. These deep forests cannot be easily reassembled once gone. Their ecological memory, accumulated over ages, is erased with each clear cut or burn.

• Mangrove Wetland: Mangroves are groves of salt tolerant trees straddling the boundary of land and sea, flourishing in tropical coastlines. Though less celebrated than rainforests, they are ecological powerhouses. Mangroves sequester carbon at rates up to ten times greater than terrestrial forests and store three to five times more carbon per acre, mostly in their waterlogged soils. Their tangled roots also provide nurseries for fish and buffer coastal communities against storms and erosion. Yet, mangroves have been disappearing rapidly. They are being cleared from shrimp farms, coastal development, and due to rising seas. By some estimates, over a third of global mangrove area was lost in the late 20th century. This destruction not only releases massive carbon stores but also leaves coastlines vulnerable. Saving remaining mangroves and restoring them is an urgent task for both climate mitigation and adaptation.

Each of these ecosystems is a linchpin in the Earth's climate and biosphere. Their potential collapse represents the unraveling of regional and even global ecological stability. Coral reefs protect shorelines and feed millions; permafrost holds the climate in its frozen grip; ocean currents steady the seasons; forests create the rain that feeds crops; mangroves shield cities from waves.

The climate emergency is not alarmism; it is a scientific fact that many natural systems are teetering on the brink. We are, in a very real sense, running out of time to pull back from these edges. The urgency cannot be overstated.

And yet, amidst this sobering assessment, there are seeds of hope and resilience. The same interdependence that makes Earth vulnerable to cascading harm also makes it capable of healing in astounding ways.

Resilience and Renewal: Learning from Nature's Healing

Nature is not passive. In many places, when given an opportunity, ecosystems have shown a remarkable capacity to bounce back, sometimes faster than expected. These stories of resilience light the way forward, reminding us that it's not too late and that our actions can make a difference. Consider the following examples of ecological renewal and innovative stewardship, which offer a counterpoint to despair:

• Whale Comeback: For centuries, great whales were hunted to near extinction, their numbers plummeting by the mid-20th century. After commercial whaling was banned in 1986, populations of humpback whales and others have rebounded spectacularly. In the South Atlantic, humpback whales that once numbered only 450 individuals in the 1950s have recovered to about 25,000 today. That's roughly 93% of their pre-whaling abundance. This quiet miracle shows nature's resilience when humans simply stop killing and start protecting. As these leviathans return, they resume their ecological roles by stirring nutrients in the oceans, fertilizing plankton with their iron-rich feces, even storing carbon in their enormous bodies. One study found that an average great whale sequesters 33 tons of CO2 over its lifetime (far more than a single tree) and that healthy whale populations boost the ocean's carbon uptake. Saving whales turns out to actually help save the climate.

• Reforestation and Rewilding: Around the globe, conscious efforts to restore forests and regenerate land are yielding fruit. Costa Rica, for example, made bold policy choices (like paying landowners to let forests regrow) and was able to double its forest cover from just

26% in 1983 to over 52% by 2019. This transformed the landscape, bringing back rainfall, wildlife, and eco-tourism, and making Costa Rica a carbon sink rather than a carbon source. In China, the Loess Plateau was once one of the most degraded, eroded landscapes on Earth but has undergone a stunning greening. A massive restoration project begun in 1999 terraced hillsides, halted overgrazing, and planted native vegetation. By 2016, vegetative cover had increased by 25%, erosion rates plummeted, and locals reported the return of birds and clear streams. What was effectively a high desert wasteland has in parts become green hills where farmers can grow crops sustainably. Even wildlife is returning as habitats recover. Meanwhile, in parts of Europe and North America, "rewilding" projects are reintroducing keystone species, like wolves, beavers, and bison, to restore natural balances. The famous reintroduction of wolves to Yellowstone Park in the 1990s, for instance, cascaded through the ecosystem. Elk over-browsing stopped, streamside vegetation regenerated, songbirds and beavers came back, and even river courses stabilized due to healthier banks.

• Innovative Conservation and Design: Human creativity, aligned with nature, is developing solutions that blend technology and ecology. In marine conservation, scientists are growing corals in underwater nurseries and replanting them on damaged reefs, boosting reef recovery. Coastal communities are restoring mangroves and marshes as "natural infrastructure" instead of building concrete seawalls, finding that green barriers better protect against storms and nurture fisheries. Farmers practicing regenerative agriculture are healing soil by treating it as a living sponge. By planting cover crops, reducing tillage, and integrating trees (agroforestry), they enhance soil structure and biodiversity, which in turn increases crop resilience to drought and sequesters carbon. Cities, too, are getting involved in regenerative design. Urban planners are creating wildlife corridors, green roofs, and biofiltration wetlands that clean water and provide refuge for birds and insects even in concrete jungles.

• Rights of Nature and Policy Shifts: Perhaps one of the most profound signs of change is not in the ecosystem itself but in our legal and moral frameworks. A movement to recognize the Rights of Nature has been growing, acknowledging legally that rivers, forests, and other non-human entities have the right to exist, flourish, and be restored. In 2008, Ecuador became the first country to enshrine the rights of Pachamama (Mother Earth) in its constitution, reflecting indigenous concepts of living well in harmony with nature. Ecuador's courts have since upheld those rights to stop projects that would harm sensitive ecosystems. New Zealand followed by granting the Whanganui River legal personhood, appointing guardians to speak for the river's interests. This was a solution born from Maori spiritual regard for the river as an ancestor. As of 2024, elements of nature have been granted legal rights in at least 34 countries, in over 300 instances worldwide. At the same time, concepts like the "planetary boundaries" framework are influencing international policy, aiming to keep human activity within safe limits for Earth's systems. The 2015 Paris Agreement on climate, though still insufficient, united almost all nations in action to limit global warming. And the 2022 UN biodiversity agreement commits countries to protect 30% of lands and oceans by 2030. These shifts in law and policy indicate a growing recognition that our own well-being is inseparable from the well-being of the Earth. They are attempts to codify an ancient truth: we must be stewards of the natural world.

Stewardship and Ecological Intelligence

The challenges we face are as much spiritual and ethical as they are scientific. We have seen that Earth behaves like a living, intelligent entity, but do we have the wisdom to respond in kind? Stewardship means recognizing our responsibility to care for the land and waters as a sacred trust. It requires humility, listening, and reciprocity.

As the ecologist and Potawatomi member Robin Wall Kimmerer writes, we must learn to be "honorable harvesters," taking only what we need and giving back in gratitude. Indigenous fire management techniques, for example, are now being adopted in western forests to prevent catastrophic wildfires, reaffirming that traditional knowledge and modern science together can guide us.

Spiritual ecology, a growing movement, calls us to rekindle a reverence for the Earth. Whether one speaks of the Earth as Gaia, as Pachamama, as God's creation, or simply as the miracle of life, the common thread is love and respect. When we see a forest not as timber but as a cathedral of life, or a river not as a resource but as the veins of the land, our relationship shifts from exploitation to care. The heart of ethical ecology is expanding our circle of moral concern to include all beings. It means asking not "what can we take?" but "what is the Earth asking of us?"

Right now, the Earth is asking for healing, protection, and balance.

It is also asking us to learn. In the face of environmental crisis, many are turning to the Earth's own systems for inspiration. Biomimicry is a field where engineers and designers emulate natural designs (like solar cells mimicking leaves, or buildings modeled on termite mounds for natural cooling). Likewise, our social systems can learn from ecosystems. Diversity, redundancy, and cooperation make ecosystems resilient; so too could our economies and communities prioritize these values. The intelligence of Earth is on display in the way mycorrhizal networks distribute nutrients to where they're needed most. If our human systems could mimic that level of distributed intelligence, ensuring resources reach the vulnerable and by recycling waste into useful input, we would create a more circular, regenerative society.

Practicing stewardship might be as intimate as tending a garden or as grand as rewilding a landscape. It might mean advocating for policies that protect the voiceless (trees, whales, future children), or simplifying one's lifestyle to tread lighter on the Earth. There is an ethical call to each of us to find our place of service in this pivotal time. As climate activist Joanna Macy often emphasizes, we are fortunate to be alive at a moment when the Earth needs us – a moment that gives our lives great meaning and urgency.

A Call to Unity and Action

Every act of conservation or compassion counters the tide of destruction; every seed planted, every stream cleaned, every emissions cut, every cultural shift towards respect, strengthens the bonds between us and the Earth.

Let us cultivate an attitude of belonging. We belong to this Earth. Its waters are our blood, its air our breath, its soil our flesh. In caring for it, we care for ourselves and our children's children. There is profound hope in this recognition of interbeing. It means we are never alone in our efforts. When thousands of communities around the world rally to restore a forest or defend a coastline, they are moved by the same life force that causes a wounded forest to sprout new green. It's life seeking life.

The task before us may seem vast. At times, it is tempting to believe we are too small to make a difference or mere sparks in the shadow of an overwhelming blaze. But even fire, in nature, is not always a force of ruin. It clears the underbrush, awakens dormant seeds, and renews the soil for life to begin again. In the same way, disruption can be an invitation to regeneration. One humble act, one clear voice, one honest decision may seem like drops in the fire, yet together they become the rising steam of transformation. What gives hope is not the absence of challenge, but the presence of choice. Earth's story is still unfolding, and we are not merely characters within it but co-authors of what comes next.

Will it be a chapter of reconciliation where humans rejoining the community of life as respectful members; or will it be a tragedy of hubris? The answer lies in the actions we choose now. The window to prevent the worst outcomes is narrow, but it is still open. Every day, every choice matters.

We can begin by listening, truly listening, to the Earth. Find places to be still. Let the wind speak. Sit quietly beneath a tree, by a stream, or in the hush of a field. Whether you return to the same spot or simply return to the act itself, this practice cultivates a quiet intimacy with the living world. Learn the names of nearby trees, notice the calls and migration of birds, feel the rhythm of the seasons in your own body. From this deep noticing grows love. And from love, the will to protect.

But presence alone is not enough. If a river is being poisoned upstream, tending your garden downstream will not save it. Personal awakening must flow into collective action and into advocacy, into policy, into the redesign of the systems that shape our shared future. True transformation requires both stillness and movement, contemplation and courage. When we root our efforts in reverence and rise together with clarity, we begin to co-create a world in which both people and planet can thrive in deep and conscious connection.

Most of all, we must keep hope alive. Hope is a verb, an active choice to trust in the capacity for change. It does not mean ignoring the challenges; it means refusing to give up on our beautiful planet. There are millions of young people, scientists, indigenous leaders, farmers, artists, and ordinary citizens already working for an ecological civilization. We are not alone, and we can draw strength from one another.

The Earth is calling. In the whisper of the trees, in the cry of an endangered whale, in the silence of a dried-up riverbed. It calls us to remember who we are. May we have the courage to respond, and in doing so, rediscover our place in the magnificent family of Earth.

Chapter 6: Consciousness – The Bridge Between Inner Life and Cosmic Unity

The image above maps our Laniakea supercluster – a vast web of galaxies including our Milky Way (marked by the red dot). It is a humbling illustration of our place in the cosmos.

"I regard consciousness as fundamental. I regard matter as derivative from consciousness. We cannot get behind consciousness. Everything that we talk about, everything that we regard as existing, postulates consciousness."

— Max Planck
Physicist and Founder of Quantum Theory
Max Planck
pioneer of quantum theory
who opened the doorway to a deeper dimension of reality
He discovered that the building blocks of matter
are not solid things
but waves of possibility
shaped by observation
and entwined with the act of knowing

For Planck
consciousness was not an afterthought
but the origin
the silent presence behind all form
His insight invites us to reconsider what is real
to see mind and matter not as separate
but as reflections of one source
In this view
the universe becomes a question asked within us
and we
the answer unfolding

Chapter Summary: This chapter explores consciousness as a bridge between subjective experience and the wider cosmos. Rather than focusing on repeated theories of quantum cognition, we enter terrain on the role of altered states, the dynamics of collective mind, and the intersection between spiritual traditions and scientific inquiry. We draw from meditative experiences, near-death encounters, and cross-cultural mysticism to examine how consciousness dissolves boundaries between self and other, time and space, form and meaning. Embracing both modern insights and ancient understandings, this chapter invites us to see consciousness not as an isolated property of brains, but as a field of connection and responsibility. The mystery of awareness becomes a call to humility, unity, and ethical stewardship of our shared reality.

Laniakea: Home Beyond Horizon

We often imagine our home as a single planet, a blue world circling one quiet star. But we belong to something far larger. The Milky Way is just one river in a vast and ancient ocean of galaxies, gently pulled together by a gravity we cannot feel.

This great structure has a name. Laniakea. It means "immeasurable heaven" in Hawaiian. It stretches across space as a gathering of light and matter too wide for any eye to see at once. We live near its edge, a red dot on a map drawn in starlight.

Laniakea is not just a place. It is a reminder. We are cradled in something immense. We are part of a structure so vast it escapes ordinary thought. The galaxies do not drift alone. They move together, pulled by a center we cannot find. So too may we be drawn toward one another, toward meaning, toward a mystery that holds us all.

Altered States and the Opening of the Inner Cosmos

Consciousness is often taken for granted like a silent witness to life's unfolding. But in altered states of awareness, it reveals itself as something vast, flexible, and deeply mysterious. In experiences ranging from deep meditation to near-death encounters and psychedelic journeys, the boundaries of ordinary perception dissolve. People report a loss of self, a merging with something greater, and an overwhelming sense of unity.

One of the most studied gateways to these states is DMT, a molecule found in both plants and the human body. Described as "the spirit molecule," DMT induces visions of kaleidoscopic light, intelligent presences, and realms that feel more real than waking life. Scientific studies show that under its influence, brain regions usually separated begin to synchronize imagination, emotion, and memory swirl into one field of perception. Mystics and shamans have long claimed that DMT allows contact with the sacred or the spiritual intelligence of nature. Whether physiological or metaphysical, these experiences challenge the assumption that consciousness is confined to the skull.

Similar insights emerge in meditation and near-death experiences (NDEs). Mediators describe states of boundless awareness in which the illusion of separation falls away. Near-death experiences often report floating beyond their bodies, encountering profound light, and merging with an all-encompassing presence. In both cases, the sense of *I* gives way to a felt truth of unity or a consciousness that seems not personal but shared. Neuroscience has even documented the softening of ego-related brain regions during such states, suggesting that the boundaries of self may be a more neurological habit than fixed reality.

What these altered states reveal is not only that consciousness is expansive, but that it is inherently relational. When the filters fall, what's left is connection to life, to others, to a reality that pulses with meaning.

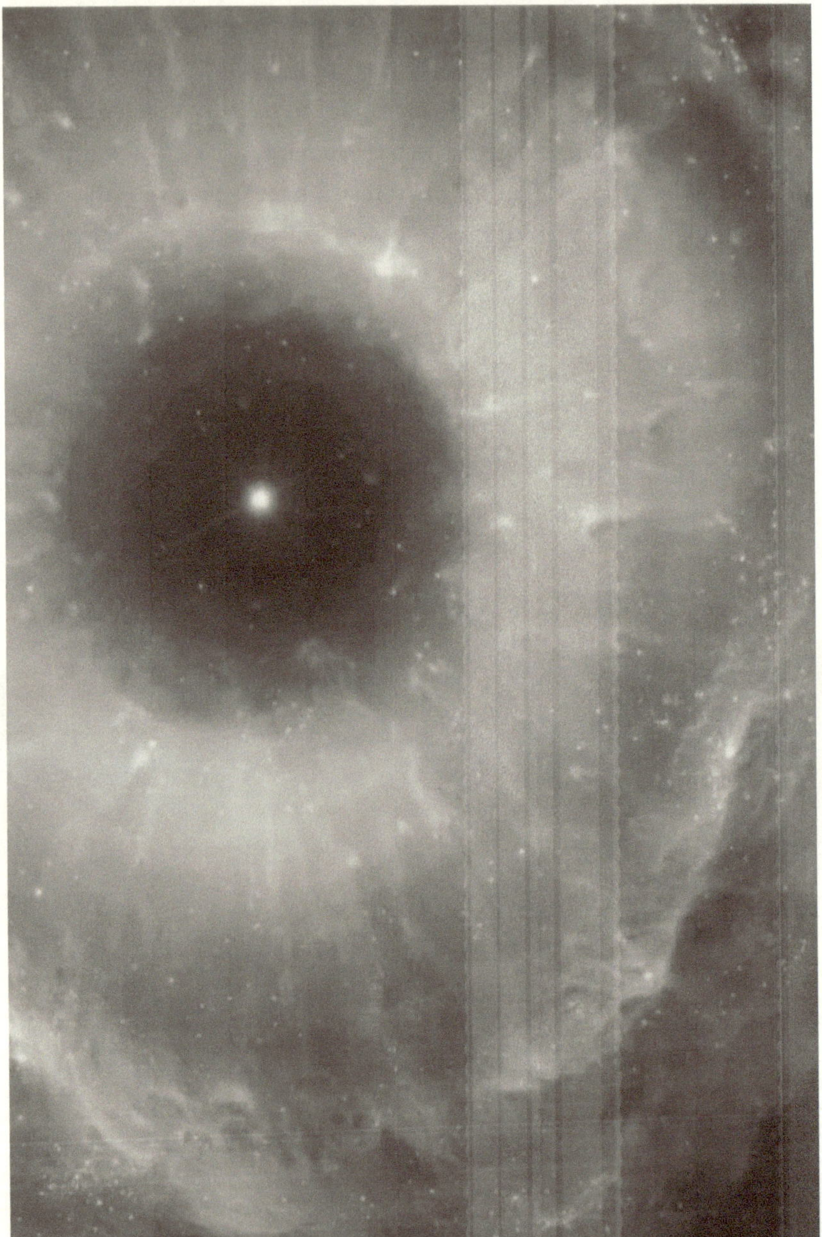

The Helix Nebula, also known as the Eye of God Nebula

◎ = The Eye That Opens Both Ways

To see the cosmos is to be seen by it

The Gaze That Watches Back

What happens when we look so deeply into the cosmos that something seems to look back?

The Helix Nebula, often called the Eye of God, is not just a spectacle of dust and gas. It is a symbol, a mirror. In its spiraling iris we see more than stellar death and rebirth. We see ourselves. Not as isolated observers, but as participants in a greater unfolding. A consciousness within consciousness.

In altered states, we often speak of thresholds. Moments when perception shifts, when the veil thins, and what once seemed external begins to speak in the language of the soul. The Eye of God becomes not only a celestial image, but an inner event. A remembering.

To investigate the vastness of space and feel known is not a hallucination, it's a resonance; a reflection of the same pattern that opens within the inner cosmos. The spiral. The eye. The self that sees and is seen.

At the edges of the mystical and the scientific, there lies a truth both intimate and infinite. Awareness is not bound to the skull. It extends, receives, and reflects. It connects.

As individuals awaken to this, something begins to stir between us. A field, subtle and shared. A mind not made of parts, but of presence. What if we are not many minds in isolation, but one mind echoing through many forms?

From the Eye of God to the eye within, we arrive at a new question. Not just *Who am I* but *Who are we becoming together?*

The Collective Mind and Cultural Fields of Awareness

Consciousness is not only an inner phenomenon. It is also social, shared, and shaped by the world around us. We think with words inherited from our culture. We feel what others feel. We absorb the stories of our time, and in doing so, participate in something larger than ourselves. We shall call this collective consciousness.

Sociologist Émile Durkheim described this as the conscience collective describing a moral and emotional field that binds societies together. Carl Jung saw it as the collective unconscious, a deep reservoir of archetypes and symbols that recur across cultures and dreams. More recently, philosopher Pierre Teilhard de Chardin imagined a thinking layer around the Earth, formed by the interwoven minds of humanity and dubbed it a Noosphere.

In the age of the internet and global networks, this noosphere is no longer theoretical. Collective grief, joy, or outrage can spread in seconds. Events like natural disasters or shared rituals tie us together. Even fringe research, such as the Global Consciousness Project, suggests that major emotional events may subtly influence random systems as though consciousness has a field-like presence that resonates on a planetary scale.

Whether metaphor or mechanism, the idea remains: our minds are not islands. Consciousness flows between us, shaped by language, history, and presence. When we awaken to this, compassion becomes more than virtue. It becomes recognition of shared being.

Mind in Nature: Indigenous Wisdom and Panpsychic Insights

While science has begun to ask whether consciousness could extend beyond the brain, many cultures never asked otherwise. In indigenous traditions around the world, consciousness is seen not as human-made but as nature-born. Forests, rivers, animals, even stones are understood to have awareness in their own way, not as metaphors, but as living presences. These traditions speak of the Earth not as an object but as a relative, a being deserving of respect and reciprocity.

This view is echoed in modern philosophical frameworks like panpsychism which is the idea that consciousness may be a fundamental property of matter itself. Rather than consciousness arising only in complex brains, perhaps it exists on a continuum, embedded in all things. Under this lens, consciousness is not rare; it is the light inside the atoms, flickering more brightly in some configurations but present everywhere.

Physicist Max Planck once said, "I regard consciousness as fundamental." Other quantum pioneers hinted that the observer and the observed are inseparable and at the heart of reality, there may be no clear line between matter and mind. This suggests that the cosmos is not a dead machine but a living mystery, and that our awareness is a reflection of that greater field.

As Western science and ancient wisdom meet, we are invited into a more unified view of existence. It is one where consciousness is not contained but continuous.

Ethical Consciousness and the Call to Wholeness

To recognize consciousness as shared, relational, and possibly universal is not merely a metaphysical idea. It is an ethical awakening. If other people are not truly "other," if nature itself is imbued with awareness, then every act becomes a moral choice.

Harming another becomes self-harm. Exploiting nature becomes a form of spiritual amnesia. Conversely, to love, to heal, to protect become not only noble but necessary. The golden rule transforms into a law of resonance: what we do to others, we do to the whole, and to ourselves.

Consciousness, then, is not a luxury to explore, but the lens through which we see what matters. And if we are all connected by it, then the most urgent task of our time is to tend that connection with care.

A Bridge Across the Infinite

To walk through the mystery of consciousness is to stand on a bridge. A bridge that spans the visible and invisible, the known and the unknowable. At one end is the singular, unrepeatable spark of your own experience. At the other is the vast unknown including the cosmos, the collective, and the sacred.

Across this bridge flows memory, meaning, joy, sorrow, creativity, and silence. And at the center of the bridge lies the place where opposites meet and where the human heart stands, aware that it is both observer and observed, both part and whole.

We may never fully explain consciousness. But perhaps that is not its purpose. Perhaps its purpose is to reveal and remind us that the universe is not mute, nor meaningless, but alive, and that we are participants in that aliveness.

Let us walk that bridge gently, and together.

Chapter 7: Unity and the Future of Human Civilization

"The destiny of humans cannot be separated from the destiny of Earth."

— Thomas Berry

Thomas Berry
cultural historian and ecological philosopher
who called himself a geologian
a voice for the living Earth
He saw as participant in a sacred story still unfolding
Through his work
he offered a new cosmology
one that wove science and spirit
time and tenderness
into a single narrative of belonging
He taught that our future
depends on rejoining the Earth community
in communion
for what we do to the Earth
we ultimately do to ourselves

Chapter Summary: In this chapter, we explore how profound ethical awakening can guide the future of human civilization toward unity and sustainability. We examine philosophical and spiritual principles of oneness, stewardship of Earth, and justice as a moral compass emerging from humanity's brush with peril. We confront the ethical dilemmas of technological ambitions, such as interplanetary colonization, and ask whether humanity is exporting its problems to the stars instead of healing them on Earth. Finally, we discuss frameworks of compassion, planetary limits, and intergenerational justice that can inform practical and visionary approaches for a humble, accountable, and globally unified civilization.

Ethics Forged from Risk

History has shown that periods of great risk often give birth to new moral frameworks. Standing at the brink of nuclear war or climate catastrophe, humanity is compelled to reflect on what truly matters. When our survival is at stake, questions of ethics, unity, and responsibility come to the forefront. The tension of existential risk; whether from technological misuse, ecological collapse, or global conflict; serves as a crucible in which a new ethic may be forged. Will we continue to act out of short-term self-interest and conquest, or will the specter of shared risk inspire a higher moral ground? The very threats that divide or terrify us can also remind us of our common destiny. In this sense, ethics emerges as a response to risk: a guiding light born from the darkness of near disaster. By acknowledging our collective vulnerability, we lay the groundwork for an ethical framework that emphasizes our shared humanity and the urgent need for unity in action and purpose.

Principles of Unity, Stewardship, and Justice: Responsibility and Neurodiversity

True unity is not sameness. Harmony does not erase uniqueness. In a living system, diversity is not a flaw. It is the source of resilience, insight, and intelligence.

When we speak of the wisdom of the whole, we must include every voice the world tries to quiet. The quiet child who avoids eye contact. The adult who moves through space differently. The mind that interprets the world through sensory patterns or abstract rhythms that others may not understand.

To recognize the needs of those with autism or other forms of neurodivergence is not an act of pity. It is an act of integrity. It is the system recognizing itself. It is the whole seeing its parts clearly and honoring their function.

In a universe shaped by pattern, some of the most important frequencies are not the loudest.

And in societies shaped by speed, conformity, and surface appearance, those who perceive differently often carry the keys to healing, innovation, and reconnection.

Design that does not account for them is not inclusive.

It is incomplete.

The responsibility is not to fix differences, but to repattern the system to include it with care, with patience, and with humility. Because if our systems do not work for the most vulnerable among us, then they do not truly work at all.

Hand in hand with unity is justice. A truly unified civilization cannot exist without justice and equity. Unity is hollow if some thrive by pushing others down. Justice means recognizing the inherent dignity and rights of all people and indeed all forms of life. It calls for fairness across nations, classes, and generations. It calls for a future of human civilization guided by unity where success would be measured not by dominance or wealth, but by how well we uplift the whole. This vision challenges us to redefine progress. It is not a race of competition, but a journey of cooperation where advances in science or technology are married to compassion and ethical reflection. In essence, unity is a spiritual principle, stewardship is its practice, and justice is its moral compass. These principles form the ethical foundation for confronting the great decisions ahead.

The Jupiter-C launched Explorer I on 31 January 1958 as the U.S.'s first artificial satellite.

Ambition Beyond Earth: Responsibility vs. Escape

The modern era has seen astounding technological ambitions, none so grand as the dream of colonizing other planets. Visionaries speak of making humanity a multi-planetary species, often citing it as insurance against extinction or a backup planet on Mars in case catastrophe strikes Earth. On the surface, this vision is inspiring: humans voyaging to new worlds, ensuring our survival by expanding into the cosmos. But beneath the excitement lie profound ethical questions. Do we have the right to escape to another planet after ravaging our own? What is our moral responsibility to Earth, our home, in the face of these ambitious dreams?

Critics of interplanetary colonization argue that rushing to plant our flag on Mars could be a form of moral evasion. It is far easier to imagine starting anew on a pristine world than to confront the messy problems we've created here. As one scientist noted, focusing on Mars can become an excuse to deflect responsibility from Earth. After all, if billions, even trillions, of dollars are poured into making Mars habitable, what message does that send to those suffering on Earth right now? Every rocket launch and life-support system for another planet carries a hidden cost in earthly resources and attention. Indeed, humanity is already consuming natural resources at ~1.7 times the Earth's capacity, living as if we had 1.7 Earths to burn. Are we willing to strip-mine our planet's remaining water, minerals, and ecosystems to fuel an escape? The ethical calculus cannot ignore that each gallon of rocket fuel, each ounce of metal, and each brain cycle spent on leaving Earth is one not spent on healing Earth.

Moreover, exporting humanity's problems to new worlds without solving them at home could simply spread the disease of unsustainability across the cosmos. If we have not solved issues of war, inequality, or environmental destruction on Earth, there is no guarantee we won't repeat those patterns elsewhere. Would a Mars colony carry humanity's baggage of greed, division, and exploitation into the heavens? This calls to mind the cautionary tales of terrestrial colonization. When European powers sailed to new continents, they often brought conquest and devastation to indigenous peoples and ecosystems. We must ask whether other planets would be different. Would Mars become a grand experiment in unity and learning from past mistakes, or a new era of imperial conquest under a cosmic guise?

Ethically, our responsibility is first and foremost to our home world. Some ethicists argue that humanity, given its poor track record of environmental stewardship, has not yet earned the privilege to colonize other planets. Our priority should be to protect the Earth before moving to another planet. Only by learning to live within Earth's means and repairing our relationship with nature can we ensure that if we one day step onto Mars, we do so with wisdom and care rather than desperation. This is not to stifle innovation or the spirit of exploration, but to reframe success not as conquest of new frontiers, but as the healing of the one garden we already inhabit.

The Hidden Costs of Colonizing Other Worlds

Digging deeper, there are specific hidden costs and ethical pitfalls in the dream of interplanetary colonization. One glaring concern is the exploitation of resources, both Earth's and those of other worlds. Space entrepreneurs discuss mining asteroids or extracting Martian minerals to support colonies. Yet, as Scientific American warns, we have seen how mining on Earth can bring devastating environmental impacts. To assume it would be gentler on Mars is wishful thinking. If anything, the harsh conditions of Mars might tempt colonists to strip mine without restraint under the justification of survival. Would Mars be treated as a blank check for resource extraction, a frontier where our worst industrial impulses run free?

There is also the ethical issue of planetary protection and potential alien life. While Mars appears barren to our rovers, we must ask: what if microbial life exists hidden beneath its surface? Do we have the right to terraform and transform Mars, possibly exterminating native life (even if only bacteria) for the sake of our own habitat? Astrophysicist Carl Sagan once argued that if Mars harbors life, however humble, then Mars belongs to the Martians, even if the Martians are only microbes, and that preserving that life should supersede any other use of Mars. This perspective challenges us to expand our circle of moral consideration beyond humanity and to test our ability to act as ethical stewards, rather than conquerors. It forces the question: Is our venture into space driven by curiosity and respect, or by a colonial mindset seeking new lands to exploit?

Even if Mars is lifeless, the act of colonization carries human risks and moral traps. One is the potential for a new kind of inequality. Imagine a future where a few wealthy pioneers live in a comfortable Martian habitat while billions back on Earth suffer the worsening climate crisis. Would off world colonies become enclaves of the privileged, a literal exit strategy for the rich? Such a scenario is no science fiction exaggeration since it often flows naturally from today's disparities. The ethics of interplanetary ambition then loops back to earthly justice. A unified civilization cannot abide a future where we flee from our failures, leaving the less fortunate behind on a deteriorating planet.

Perhaps the hidden cost most difficult to quantify is opportunity cost. Every scientist, engineer, or dollar devoted to Mars is one less applied to sustainable energy, poverty alleviation, or peacebuilding on Earth. Yes, space exploration yields innovations, but we must balance that against what we choose not to do here and now. There is a poignant symbolism in the idea of searching for water on Mars while millions on Earth lack clean water; or trying to grow food in Martian soil while our fertile lands crumble from mismanagement. Are our dreams of the stars blinding us to our responsibilities under our own sky? The ethical challenge is to dream big without escaping the obligations of the present. If we do venture outward, it should be with humility, ensuring our home planet and all its inhabitants are not sacrificed at the altar of ambition.

Ethics for a Planetary Civilization

If humanity is to have a future worth celebrating, whether on one planet or many, it must develop ethics for planetary civilizations. This means an ethical framework expansive enough to account for global interdependence, technological power, and our duty to future generations. Several complementary principles offer guidance: compassion, planetary limits, and intergenerational justice.

Compassion is the simple yet profound principle of empathy in action. It calls us to care for others' well-being as if it were our own. In a unified civilization, compassion would not be relegated to personal charity; it would be embedded in our policies, economies, and technologies. A compassionate lens asks: who is affected by our choices, and how might we minimize suffering? For example, decisions about energy or industry would center on care for the vulnerable and for other species, not just profit or convenience. Compassion reminds even the most brilliant minds that no achievement is truly great if it leaves misery in its wake. As environmentalist Gus Speth observed, our greatest problems may not be technological at all, but human: "the top environmental problems are selfishness, greed, and apathy," and overcoming those requires a "spiritual and cultural transformation" beyond what science alone can provide.

Planetary limits refer to the recognition that Earth is finite, and we must live within its means. Scientists have identified planetary boundaries or thresholds in Earth's life support systems that we transgress at our peril. An ethical civilization takes these limits seriously as guideposts for sustainable living. This means embracing an economy and lifestyle that fits Earth's budget of resources and waste absorption. Just as an engineer respects the limits of materials, society must respect the ecological limits of the planet. We might invoke the old adage, "live simply so that others (and other species) may simply live." By honoring planetary limits, we demonstrate humility by acknowledging that humanity is not above nature's rules and in foresight by ensuring we don't undermine the very foundations of our survival. This principle directly challenges the notion that success equals endless expansion. Instead, it frames success as achieving balance and thriving human cultures in harmony with the Earth's rhythms.

Intergenerational justice extends moral concern to those who will inherit the consequences of our actions and the future generations of humans (and even the future of other life). It asks us to consider: how will our choices today affect the children of tomorrow, and do we have the right to compromise their world for present gain? This concept has gained real traction; many nations now recognize the rights of future generations in climate laws. The idea is often captured in a simple, powerful mandate: leave the world better (or at least no worse) than you found it. In practice, intergenerational justice might be used as a litmus test for policy. For instance, any plan for development or technology should be evaluated for its impact on people decades or centuries from now. The climate crisis has made this clear: a short-sighted pursuit of growth and comfort by earlier generations has put today's youth and unborn descendants at risk in a diminished world. We are challenged now to break that pattern. As one moral maxim states, "we do not inherit the Earth from our ancestors; we borrow it from our children." Justice demands that we return it to them intact and flourishing. This framework promotes long term thinking and guardianship so we may see ourselves as trustees of a legacy.

When we combine compassion, planetary limits, and intergenerational justice, we begin to sketch an ethical blueprint for a planetary civilization. These frameworks encourage decisions based on care and respect for all life (present and future) within the Earth's carrying capacity. They also help navigate complex dilemmas. For example, consider geoengineering by intentionally altering Earth's climate to mitigate warming. A purely technical mindset might charge ahead with a grand solar shield or aerosol plan. But ethical frameworks would urge caution: compassion asks who might be harmed by side effects (perhaps vulnerable communities); respect for planetary limits asks what natural systems we might disrupt; intergenerational justice asks what burden or risks we shift to those after us. Such considerations civilize progress. They ensure that as our knowledge and power grow, so too does our wisdom and restraint.

Building a Unified, Equitable, and Sustainable Civilization

Importantly, building a unified civilization does not mean a uniform civilization. We are not striving for a monolithic global culture where differences are erased. Unity, in truth, celebrates diversity while aligning around common values of respect and care. Just as an ecosystem is robust when it has diverse species playing different roles, a global human society is stronger when it preserves a mosaic of cultures, languages, and perspectives, provided they are in dialogue and share a commitment to the well being of the whole. Therefore, part of the project is creating forums and norms for intercultural understanding, so that our rich differences become a source of creativity rather than conflict. The principle of justice again comes in: historically marginalized voices must be included and heard in shaping the future. The wisdom of indigenous peoples, often ignored, offers keys to living in balance; the experiences of the poor and vulnerable illuminate what real needs are; the insights of philosophers and scientists together can guide ethical innovation. In sum, a unified, sustainable civilization will be an ongoing conversation which is as inclusive and compassionate as possible.

Power, Influence, and the Need for Humility

Any discussion of our future would be incomplete without a critique of power, influence, and ego. A handful of individuals and nations wield unprecedented influence over the course of civilization. Tech billionaires launch private spaceflights; great powers set policies that affect billions; charismatic leaders inspire movements or, alternatively, stoke divisions. Power itself is not inherently evil since it can achieve great good, but how it is wielded depends on the character and wisdom of those who hold it. As Abraham Lincoln once noted, "Nearly all men can stand adversity, but if you want to test a man's character, give him power." And indeed, the test of character is upon us collectively: will the power we've amassed as a species (through technology, knowledge, wealth) be guided by ego or by humility?

Ego is a double-edged sword. On one side, ambition and confidence, the byproducts of ego, have driven humans to innovate and explore. On the other hand, ego unchecked leads to arrogance, domination, and blindness to one's own limitations. When we look at endeavors like interplanetary colonization or geoengineering, we must be wary of ego masquerading as vision. The line between a bold solution and a hubristic folly is thin. A thoughtful critique of influence asks: are our leaders and innovators truly listening to feedback, to dissenting voices, to the lessons of history? Or are they charging forward fueled by a conviction of their personal rightness and glory? The difference is critical. A leader guided by humility remains open to counsel, acknowledges uncertainty, and prioritizes the common good over their own legacy. A leader driven by ego may ignore warnings and plow ahead to be the first, the biggest, or the most famous, regardless of the consequences.

In shaping the future of civilization, we as a society need to demand humility from ourselves and from those in positions of power. This means cultivating a culture that does not equate admission of error with weakness, nor patience and caution with lack of vision. True humility is not timidity; it is honesty about our place in the world and the impact of our actions. It is the quality that tempers power with conscience. One could say it is the antidote to Icarus syndrome: flying too high on the wings of pride. If our species is Icarus, humility is the recognition that the sun's heat (nature's laws) demands we fly at a safe height. Humility teaches us to ask, *"Should we?"*, not just *"Can we?"*

It reminds us that no individual, no matter how brilliant, sees the full picture alone. In a sense, humility is a collective intelligence and allows room for many voices and course correction when needed.

Consider the ego-driven narrative of conquest versus a humility driven narrative of care. The conquest narrative celebrates feats like conquering Everest, conquering disease, or conquering new markets. In contrast, the care narrative would speak of learning from Everest (and protecting its sacred slopes), healing disease, or serving new communities. The language shift is subtle but profound. It signals a move from dominance to partnership. The future of human civilization might depend on replacing the conquering hero image with that of the wise caretaker. It means valuing leaders and everyday heroes who are healers, bridge builders, and teachers. It means celebrating cooperation and compassion as greater accomplishments.

We should also be honest about influence and accountability. In a globally connected civilization, the actions of one nation or corporation can affect the entire biosphere. Power must come with accountability at a global scale. This could entail stronger international laws to check abuses like environmental damage or exploits that could jeopardize humanity (like unsafe AI development or biological research without oversight). It also means that citizens worldwide need to engage and keep power holders honest. The antidote to ego driven decision making is transparency and participatory governance. A unified civilization is not one where a single entity calls the shots; it's one where collective decisions emerge from dialogue, respect, and informed consent of the governed. Achieving this won't be easy and current systems are far from this ideal, but the awareness of the problem is the first step.

Reflections and the Road Ahead

As we conclude this chapter and look toward the final message of *The Quantum Call to Unity*, it is time to turn the questions to the reader, to you. The future of human civilization is not an abstract concept or someone else's responsibility. It is being shaped now by each of us, by our values and actions large and small. So, what does unity mean to you, not in theory but in practice? How might you live differently if you truly embraced the idea that your fate is intertwined with that of a farmer in a distant land, a child yet unborn, and even the bees in your garden? Would success still be defined by what you personally achieve, or rather by what we accomplish together for the greater good?

Take a moment to imagine the world a century from now. Will it be a multiplanetary society thriving in harmony, having solved the riddles of sustainable living and ethical expansion? Or will it be a cautionary tale of a species that reached for the stars while neglecting the gentle blue cradle that birthed it? Perhaps it will be something in between but the outcome is not written in the stars; it is written in our choices. Ask yourself: Are we doing enough to shift the trajectory? Are our definitions of progress and success aligned with a survivable, even beautiful future?

Finally, consider the role of humility and care in your own life's aspirations. The narrative of conquest is seductive: to be the best, to be the first, to accumulate and to triumph. But zoom out and view our Pale Blue Dot, as Carl Sagan did: a speck of dust suspended in a sunbeam, home to everyone you love and every human drama ever played out. In that light, might there be greater heroism in compassion than in competition? In the end, which will matter more: the monuments we build, or the lives we touch and the love we cultivate?

These are not easy questions, but they are liberating ones. They free us from the smallness of ego and invite us into the grandeur of unity. The call to unity, this quantum call that reverberates at every level from the subatomic to the cosmic, asks us to rethink what it means to be human and what future we aspire to create. As we move into the final part of this journey, carry these reflections with you. The story of the future is still being written, and every one of us holds a loomquill.

Will we write a story of humility, accountability, and global belonging where success is measured by how well we care for one another and our planetary home? The answer, dear reader, is waiting to be lived. Let us proceed with hearts open and minds united, toward a future where unity is not just a distant ideal, but the lived reality of human civilization.

Chapter 8: Restoration and Healing

"It is a privilege to be alive in this time when we can choose to take part in the self-healing of our world."

– *Joanna Macy*

Joanna Macy
teacher of deep ecology and the Work That Reconnects
whose life has been a bridge between grief and empowerment
She invites us to feel the full weight of the world's sorrow
not as a burden
but as a gateway to transformation
Her voice reminds us that despair is not the end
but a sacred turning point
when the heart breaks open
to the call of interbeing
Through her guidance
countless people have remembered
that active hope is a choice
and that by showing up fully
we become part of the Earth's healing process
alive to the wonder of this moment
and the responsibility it carries

Chapter Summary: This chapter explores restoration as an active, ongoing process of healing. Ultimately, Restoration offers a hopeful vision of growth through repair, a journey of mending our world, our communities, and our souls in unison.

Earth and moon as seen from Space. Credit: NASA/NOAA/DSCOVR

Healing the Self: Transcending Trauma and Finding Wholeness

The journey of restoration also runs within each of us. Every individual carries their own wounds of personal losses, childhood traumas, broken hearts, and existential anxieties. Healing the self is a deeply personal process, yet it is profoundly connected to Earth and society. As we heal internally, we become more capable of nurturing others and the world; and as we witness healing in nature and community, we find strength and hope for our own journeys. Personal healing is not about forgetting pain or returning to some naive past innocence. Personal healing is an ongoing, active engagement with our hurt, transforming it into wisdom, compassion, and growth.

Modern psychology has shown that humans have an innate capacity for resilience and even transformation after trauma. The concept of post-traumatic growth describes how many people emerge from hardship with new insights, deeper relationships, and a renewed sense of life's meaning. But such growth does not happen automatically; it unfolds when we face our pain and work through it, often with support. Practices like mindfulness and meditation have become widely embraced as tools for this inner work. By learning to sit with our thoughts and feelings without judgment, we gradually release their grip. A person haunted by anxiety, for instance, may find that mindful breathing and grounding techniques create a space of safety in the present moment. Over time, the hyper-vigilant nervous system can relax, re-learning that not all moments are the trauma. Neuroscience confirms that mindfulness can actually rewire neural pathways, quieting the brain's fear centers and strengthening areas that regulate emotion and foster empathy. In essence, mindfulness helps heal the mind-body connection frayed by trauma, teaching our whole being that it is safe to experience peace again.

Another pillar of self-restoration is reconnection with meaning and the sacred. Trauma and depression often involve a feeling of disconnection whether from others, from purpose, or from the divine (in whatever form one conceives it). Healing invites us to reconnect those threads. This can happen through therapy and introspection, but also through experiences that awaken awe and gratitude. Research in positive psychology finds that experiences of awe like beholding a magnificent sunset, standing under the Northern Lights, or listening to moving music can shrink the ego's chatter and make us feel part of something larger and wondrous. Likewise, gratitude practices, such as keeping a journal of things one is thankful for each day, have been shown to improve mood and even physical health. Gratitude shifts our focus from what's wrong to what has helped us, reinforcing a narrative of support and hope even in hard times. Many spiritual traditions, old and new, emphasize gratitude and wonder as keys to healing, whether it's saying a prayer of thanks, offering songs of praise, or simply bowing one's head in appreciation of life's small gifts. Cultivating these feelings isn't about denying suffering; rather, it's about balancing the darkness with light, remembering that even amid pain, life offers moments of beauty, love, and grace.

Grace is the notion of unearned blessing or forgiveness. It can be a profound element in personal healing. Sometimes, healing comes through surrender. Surrendering means accepting help, letting go of self blame, or opening ourselves to forgiveness (of ourselves or others). For a person wracked by guilt or shame, the realization that they are still worthy of love and redemption can be life-changing. It's like a cracked vessel suddenly filled with gold. In fact, the Japanese art of Kintsugi offers a beautiful metaphor. When a pottery bowl breaks, the Kintsugi craftsman mends it with lacquer mixed with gold powder, treating the breakage as part of the object's history and design rather than something to conceal. The repaired bowl, with its veins of gold glinting through the cracks, is often more exquisite than the original. So it is with us: our scars and imperfections can become sources of strength and beauty. Many survivors of trauma find that in helping others with similar wounds, their own scars take on meaning and even pride. It becomes the mark of someone who has been broken and made whole again, with a more open heart. As the poet Rumi wrote, *"The wound is the place where the Light enters you."* In healing, our very wounds can become gateways to deeper light within ourselves.

Personal healing is greatly aided by community and storytelling. No one heals alone. Whether through a trusted friend, a support group, a counselor, therapist, psychologist or family elder, sharing our story is often the first step to reclaiming it. When we tell the story of our pain out loud and someone truly listens, bearing witness without judgment, an alchemy occurs. The tangled, unspoken grief or trauma starts to take a coherent shape, a narrative with a beginning, middle, and the possibility of an end. Psychologists have found that giving voice to trauma (in therapy or journals) helps integrate the experience, moving it from raw sensation to processed memory. In many indigenous cultures, healing ceremonies often center on storytelling, song, and ritual enactment of suffering and relief. These are practices modern trauma therapists increasingly acknowledge as powerful means of release and integration. Something as simple as a communal circle where each person shares one thing they are struggling with and others offer their support can be deeply healing. It affirms that we are not alone in our struggles and that our life experiences, even the painful ones, matter and carry lessons. Likewise, listening to others' stories can heal us by fostering empathy and perspective. When we hear how someone else walked through fire and came out the other side, we may discover courage for our own path.

Reconnecting with the natural world is another intimate way individuals heal. Ecotherapy, or nature based therapy, is founded on the idea that being in green, living environments restores our mental and emotional balance. A walk in the woods, gardening, sitting by the ocean, even gazing at a single tree have measurable calming effects on our physiology by reducing blood pressure, lowering stress hormones, and improving mood. But beyond the biochemical, there's a soulful element when nature often provides us with metaphors. The sight of a tree that has grown around a boulder, or a flower blooming from a crack in concrete, can remind us that life finds a way to move forward, adapting and thriving despite obstacles. Many people report that in moments of deep despair, turning to the wilderness or even caring for a pet helped them find solace when human company could not. In those moments, the boundary between self and world softens when one can feel held by something larger, whether we call it Mother Earth, God, or the spirit of life itself. This sense of being held and connected is the opposite of trauma's isolation. It's why so many spiritual traditions send seekers into nature (to the desert, the mountain, the forest) for insight and healing. In the stillness of the wild, we often confront our true self and also feel the embrace of creation, reminding us that we belong here in the family of all beings.

Wholeness Through Weaving the Strands Together

Healing the Earth, healing society, and healing ourselves are not separate endeavors, but three strands of one braided cord. Progress in one strand strengthens the others. When we engage in restoring natural ecosystems, we often find that we ourselves feel restored by the very act. Planting trees or saving wildlife can imbue us with a sense of purpose and connection that heals our inner despair. It can also bring communities together, bridging social divides as people unite for a common Earth care goal. Conversely, as individuals and communities heal their trauma and learn to treat each other with empathy, they tend to live more gently on the Earth by consuming less in pursuit of false comfort, caring more about the future of other species, and seeing nature as part of their family. In this way, inner healing and social healing contribute directly to ecological healing. A society that has learned compassion and balance will not ravage the planet without regard; it will seek harmony. And a person who finds meaning and wholeness within is more likely to extend love outward to neighbors and to other creatures.

This profound interconnection has always existed, but we are coming to understand it more clearly. Some speak of a burgeoning field of eco-psychology that treats the emotional estrangement from nature as a key factor in human suffering. Likewise, environmentalists now often acknowledge that you cannot ask people to care about saving forests or coral reefs if those people are struggling in social misery or personal trauma; their immediate pain will overshadow distant concerns. Thus, to truly heal our planet, we must also heal the societal injustices and individual wounds that keep us locked in patterns of short-term survival mode or destructive coping. Healing has to happen holistically. It is all one healing.

Importantly, restoration is not about returning to some bygone state of perfection. Rather, it is an ongoing becoming. The Earth of the future will not be the same as the Earth of the past, but it can be lush and abundant in new ways. A healed society will not erase all scars of history, but it can carry them with honor and learn from them to avoid repeating harm. A healed person doesn't forget the injury they lived through; instead, they weave it into the narrative of their life as a source of empathy and strength. We are called now to a kind of global kintsugi project: to face the brokenness around and within us, and mend it with compassion and wisdom.

Throughout this chapter, and this journey, runs a golden thread of hope and purpose. It is the recognition that the very act of working toward restoration, however imperfect and long the road, and gives our existence profound meaning. In healing what is broken, we answer a deep human longing to repair, to set things right. Many spiritual traditions suggest that humans are *co-creators* or *partners* in creation. Perhaps our role is indeed to be the healers of the web of life, to use our unique capacities to mend. When we engage in this work, we often feel a sense of alignment with something greater than ourselves. Grace descends when we act out of love for life.

Consider the moments that most move us: neighbors forming a human chain to rescue someone from rising floodwaters, rescuers working tirelessly through the night to reach children trapped in a collapsed mine, a stranger kneeling beside another in quiet solidarity, offering nothing but presence and an open hand. In such moments, we see glimpses of a world made whole. We experience the awe of unity. Gratitude naturally arises. Gratitude for each other, for the chance to help, for the simple breath that allows us to continue. And that gratitude fuels further healing action, in a benevolent cycle.

In the end, *Restoration – Healing* is about coming home to wholeness. It is remembering that we belong to Earth, to each other, and to ourselves in a sacred way. Healing is how we acknowledge that belonging. It is how we tend to the fractures so that life's unified light can shine in, through, and out. Each of us, in our own sphere, can engage in this holy work of restoration: tending a garden or a relationship, speaking truth to injustice or gently to our own wounded heart, planting a seed or planting an idea of hope in someone's mind. No act is too small, because everywhere we do this, life grows. As we heal, we participate in the great story of the world healing itself. And as Joanna Macy so beautifully said, it is our privilege to be alive in this time, when such meaningful work is calling to us.

Let us answer that call, the quantum call to unity through restoration, with courage and with joy. In mending what is broken, we become whole. In restoring our Earth, our human family, and our souls, we discover that all along, we are instruments of one great healing, and a deeper harmony that binds all life together.

Chapter 9: The Cosmic Tapestry of Unity

"The total number of minds in the universe is one. In fact, consciousness is a singularity phasing within all beings."

— *Erwin Schrödinger, Physicist and Nobel Laureate*

Erwin Schrödinger
a physicist whose inquiries reached far beyond the laboratory
He is remembered for shaping quantum mechanics
but also for asking timeless questions
about life, mind, and meaning
In this vision
consciousness is not divided by skin or species
it is a single awareness
shining through many forms
He drew from both science and Vedanta
blurring the line between physics and philosophy
His insight echoes across traditions
reminding us that the self we call "I"
may be the same light
seen through countless windows
and the universe
a great remembering
of its own eternal being

In this final chapter, we stand at the confluence of quantum physics, higher-dimensional geometry, consciousness studies, and metaphysical insight. Our journey has led us through the natural world and human understanding, and now we weave these threads together into a single tapestry. The Quantum Call to Unity echoes through every dimension of reality, inviting us to view existence from a higher vantage. By integrating perspectives from 4D spacetime to 10D string theory frameworks, from the subatomic quantum realm to the reaches of human awareness, we begin to perceive an underlying wholeness.

What emerges is a profound recognition: at the deepest level, all is one, unified across space, time, and scale.

Higher-Dimensional Perspectives: Seeing Beyond 3D

Let's explore the mystical edge where physics, consciousness, and metaphysics converge. From a higher-dimensional point of view, our familiar world of three dimensions is revealed as only a slice of a larger reality. Just as a flat 2D creature would perceive only cross-sections of a 3D object, our 3D minds perceive only fragments of higher-dimensional forms. Consider the concept of the tesseract, a four-dimensional hypercube. We cannot fully imagine a 4D shape, but we can observe its shadow in our world.

A 3D projection of a tesseract, or four-dimensional hypercube, casting its radiant shadow into our three-dimensional world.

A Glimpse Beyond Dimensions

As the cube folds inward and outward, we are reminded that what we call space is not the limit of form, only the limit of perception. The tesseract can be unfolded into eight cubes in 3D space (just as a cube unfolds into six squares in 2D), illustrating how higher dimensions encapsulate and extend lower ones. If a 4D being were to lift an object off a flat table, the 2D inhabitants of that table would see the object vanish inexplicably. Likewise, phenomena that seem "beyond explanation" in 3D may be perfectly natural when seen from a higher dimension.

One striking implication of a fourth spatial dimension is the treatment of time. In relativity, time is already woven with space into 4D spacetime, a union where space and time are two facets of one continuum. A higher-dimensional perspective takes this further: what we experience as time-bound sequence might be seen all at once from the 4D viewpoint. In analogy, a three-dimensional observer sees an entire landscape painting (past, present, future of the 2D world) in one glance, while the 2D beings in the painting experience it line by line. From a 4D frame, one could conceivably view every moment of our 3D timeline laid out before them. This idea echoes the relativistic insight that the distinction between past, present, and future may be an illusion of limited perspective. As Minkowski, Einstein's teacher, famously proclaimed: space by itself and time by itself are mere shadows – only their union holds independent reality.

Such higher-dimensional thinking provides a unifying vision. It suggests that many mysteries, from quantum "spooky action" to cosmic coherence, might be understood as geometry and connections in unseen dimensions. We begin to sense that what appears disconnected or mysterious in our 3D world could be simple and interconnected when viewed in 4D or beyond. Our universe, in fact, may be the 3D "shadow" of a higher-dimensional reality, with patterns that only fully make sense when seen from that lofty perspective.

The Observer and the Observed: Quantum Connections

Quantum physics has taught us that the role of the observer is central. At the subatomic level, the act of observation isn't a passive glance. It actively shapes reality. In the famous double-slit experiment, an electron or photon behaves like a spread-out wave until it is measured; at the moment of observation, it "collapses" into a particle at a particular location. This observer effect blurs the line between subject and object, hinting that consciousness or measurement is woven into the fabric of what we call reality. As physicist John Wheeler suggested, we live in a "participatory universe" where observer and observed form a single system.

Perhaps, from a higher-dimensional or holistic view, this makes perfect sense: the observer and the observed are not truly separate but are facets of one underlying whole. Quantum entanglement has confirmed that this effect is real and cannot be explained by any classical separation or hidden signals. Entangled particles behave as if they are one, unified across space and time. This is a fact that deeply puzzles us in 3D, yet might be simply the "shadow" of unity visible from 4D. In a higher-dimensional framework, all points in space and time could be connected through unseen links, so that what we call entanglement is just a manifestation of an underlying wholeness.

Quantum theory has thus revealed a profound interconnectedness. The classic notion that the world can be broken into independently existing pieces gives way to a picture in which relationships define reality. As physicist Fritjof Capra observed decades ago, quantum theory reveals a basic oneness of the universe where the parts have meaning only within the larger context of the whole. Likewise, physicist David Bohm spoke of an "implicate order" in which all things are enfolded together, and what we see as separate is only the explicate (unfolded) order. Modern interpretations like Rovelli's relational quantum mechanics even propose that properties of particles (like having a definite position) exist only relative to other systems. It's an ontology of connection rather than isolation.

Threads of Unification in Physics: Dimensions and Forces

Einstein's general relativity showed that space and time are unified, and that gravity is not a mysterious force acting at a distance, but the curvature of spacetime itself. This geometric unification was a revelatory step. It meant that what we experience as gravitational attraction is actually shape and connection, the fabric of the cosmos guiding matter along its curves.

Inspired by this success, Theodor Kaluza in 1919 took unity a step further. Kaluza hypothesized that if we extend Einstein's 4D spacetime with an additional, tiny curled-up dimension, we can merge gravity and light. In this 5D geometry, Einstein's equations naturally give rise to Maxwell's equations of electromagnetism. In other words, the effect of gravity leaking into that hidden fifth dimension would appear to 3D beings as an electromagnetic force. Oskar Klein soon refined this idea by suggesting the extra dimension is compact (rolled up in a circle of microscopic size), explaining why we don't see it in everyday life. Although Kaluza-Klein theory was initially regarded as odd, it achieved the remarkable feat of unifying two fundamental forces of nature and hints that higher dimensions could be the key to a Theory of Everything.

Decades later, physicists did indeed find higher dimensions indispensable in the quest for unity. String theory emerged in the late 20th century, proposing that the elementary particles are not point-like dots but tiny vibrating strings of energy. In order for the mathematics of string theory to be self-consistent, these strings must vibrate in a space of ten dimensions, (nine of space and one of time), an abstract but compelling hint that our universe might have hidden dimensions curled up beyond our perception. In these models, all four fundamental forces (gravity, electromagnetism, and the two nuclear forces) can potentially unify as different manifestations of one fundamental entity (strings or membranes) resonating in a higher-dimensional space. While experimental proof remains elusive, the elegance of the idea is profound: all the diversity of particles and forces might boil down to the music of a single super-dimensional instrument.

Parallel developments like M-theory (an extension of string theory) suggest even 11 dimensions, and brane-world scenarios imagine our 3D universe as a membrane floating in a higher-dimensional "bulk." In such views, gravity might spread through extra dimensions (explaining why it is weaker than other forces), or perhaps whole other universes could be a hair's breadth away in the higher-dimensional bulk, invisible to us.

What these theories share is a unifying spirit. They extend the fabric of reality by adding dimensions in order to sew together forces that once seemed distinct. Each added dimension is like an added thread that, when woven in the right way, ties the physical laws into a more integrated tapestry. The ultimate dream is a single framework that embraces the very large (cosmic gravity) and the very small (quantum particles). Every indication, from black hole physics to cosmic observations, suggests that, at extreme conditions, these realms intersect. As researchers at Quantum Gravity Research put it, a true theory of everything must unite general relativity with quantum mechanics, and intriguingly, some believe the golden ratio (Φ) might sit at that nexus. Indeed, the golden ratio has appeared in equations related to black hole thermodynamics and quantum gravity's fundamental constants, hinting that nature's deepest workings have a hidden symmetry or harmony.

Sacred Geometry and the Golden Thread of Growth

Throughout nature and physics, certain patterns and numbers emerge again and again, as if reflecting a cosmic design. As previously mentioned, the golden ratio, $\Phi \approx 1.618$, is a number that has fascinated thinkers from ancient Pythagoreans to modern scientists. Φ appears in the spirals of sunflowers and nautilus shells, the branching of trees, the geometry of crystals, and even in the spiral arms of galaxies. This sacred geometry is not merely aesthetic; it may be woven into the fabric of reality at all levels. Some recent theories boldly suggest that the golden ratio could be a fundamental constant of nature, appearing everywhere from quantum to celestial scales". Remarkably, Φ shows up in the mathematics of black holes and quantum states in certain black hole energy levels and even in the probabilities of quantum matrices. Such findings imply a deep link between the structure of spacetime, the behavior of matter-energy, and the mathematics of harmony.

Why the golden ratio? One clue is that Φ is intimately connected to the Fibonacci sequence and the logarithmic spiral . A sunflower packs its seeds in spirals of 55 and 89 (consecutive Fibonacci numbers) to maximize packing efficiency; the ratio of successive Fibonacci numbers oscillates around 1.618.

The golden ratio's ubiquity speaks to a unity of design. Whether in the spiral of our DNA or the whirlpool of a galaxy, we find a common mathematical thread. The ancient idea of a "music of the spheres," a harmony underlying existence, gains a modern interpretation through these patterns. Φ also often appears alongside *e*, the base of natural logarithms (~2.718), which governs continuous growth and decay (as in population growth, compound interest, or radioactive decay). Φ and *e* together could be seen as complementary symbols of life's unfolding where one (Φ) is structuring form in an optimal way, and the other (*e*) is driving processes dynamically. In the grand symphony of the cosmos, they are like two notes that, when played together, produce the chord of natural growth.

The Integration Equation: A Sigil of Wholeness

At this culmination of our exploration, it is fitting to present a symbolic synthesis of these ideas. This is the equation of unity in an expression:

$$\Xi = \int \psi(x, y, z, t) \times \Phi\text{^}e \, dt \to \infty$$

Where:

• Ξ = the integrated whole or higher-dimensional conscious perspective

• ψ(x, y, z, t) = consciousness as a wave function across space and time

• Φ = the golden ratio (approximately 1.618)

• e = Euler's number (approximately 2.718), representing natural growth and emergence

• Φ^e = harmonized growth and pattern

• ∫ ... dt → ∞ = integration over all of time, signifying eternal unfolding and unity across temporal dimensions

This compact formula is less a literal scientific equation than a sigil: It is a symbolic representation condensing our journey's insights. Let us unpack its meaning. Here, ψ(x,y,z,t) evokes the quantum wave function, a function of space (x,y,z) and time (t), which in physics represents the state of a system or the realm of potentiality. We use ψ here to symbolize the field of consciousness permeating space and time and the idea that consciousness (or awareness, or information) has a wave-like presence diffused throughout the cosmos, much like a quantum field. Multiplying this by Φ^e inserts the golden ratio (Φ) raised to the power *e* (the base of natural growth). Φ^e is a curious combination: the golden ratio, emblem of organic harmony, enhanced exponentially. In the sigil, Φ^e stands for harmonized growth which is the integration of the Fibonacci spirals of life with the continuous expansion of spirit. It links the geometrical growth pattern (Φ) with the analytical growth constant (*e*), bridging intuitive design with mathematical progression.

The integral sign ∫ instructs us to integrate over time (t) from our present into eternity (as indicated by the arrow to infinity). In calculus, integration is essentially summing up or accumulating contributions. Here it signifies summing over all time including past, present, and future for all of history and all of destiny. We integrate ψ × Φ^e with respect to time, meaning we accumulate the conscious potential (ψ) harmonized by natural growth (Φ^e) across the entire stretch of existence. The result is denoted by Ξ (Xi), a letter chosen to represent the totality or the ultimate state of integrated consciousness. And as the equation suggests, this integral tends toward infinity, implying that the potential when you integrate consciousness and harmony over all space-time is boundless. In poetic terms, Ξ could be thought of as the Omega Point of consciousness or the infinite culmination where all awareness converges into One.

While this equation is symbolic, it resonates with real concepts. The idea of integrating over time echoes the block universe view in relativity (where past and future are equally part of a fixed 4D spacetime) but here we allow for an open-ended future arrow (→ ∞) symbolizing continual growth. The presence of Φ and *e* together hints that the universe's unfolding may be guided by principles of self-similarity and exponential creativity; perhaps reflecting how life evolves, and consciousness expands. In a way, we are summing the contributions of every moment of insight, every act of compassion, every quantum interaction, every growth spiral in nature, blending them into an ever growing whole. The arrow to infinity suggests that this process of integration is endless since there is always more to discover, more to become. Unity is not a static state but a living, dynamic process, ever approaching an infinite fullness.

Seen as a cosmic sigil, Ξ = ∫ ψ × Φ^e dt → ∞ is a one line prayer of unity between mind, matter, mathematics, and spirit. It reminds us that when consciousness (ψ) operates in harmony (Φ^e) over the fullness of time, the result is an unbounded ascension toward wholeness (Ξ). It is an invitation as much as a description. *The Quantum Call to Unity* is an invitation to each of us to participate in this integral process and to add our own consciousness into the grand integral of the cosmos.

Epilogue: The Quantum Call to Unity

Standing on the threshold of this understanding, we find ourselves looking at the world with new eyes. The insights from quantum physics, higher dimensions, and the nature of consciousness all point toward a profound unity.

This realization carries both an intellectual and a spiritual significance. Intellectually, it suggests that the ultimate laws of nature are holistic and that a final Theory of Everything will not fragment the world into parts, but show it as the expression of a single principle or essence. Whether through geometry, algebra, or something even more exotic, the trend is toward unification: space and time unified, forces unified, perhaps matter and consciousness unified. The symbolic equation Ξ hints that even mathematics and meaning might join hands. If this is the direction of fundamental science, then knowledge itself is converging on the perennial wisdom taught by sages that behind the multiplicity of forms lies a single formless source.

Spiritually, the call to unity asks us to feel this truth in our hearts and live it in our lives. Higher-dimensional awareness is not only a physics concept; it is a metaphor for expanding our consciousness beyond the small self. We are invited to step into a 4D mindset in order to see time bound challenges from the timeless perspective, to recognize that the differences between people or species are like the different faces of a multidimensional jewel. When we shift into this awareness of unity, compassion flows naturally. If everything and everyone is connected with me at some fundamental level, then to love others is truly to love an extended version of myself. Hurt done to another is hurt to the whole of which I am a part. In the quantum hologram of the universe, each part contains the whole; thus our choices ripple outward in ways we cannot fathom, but which a higher-dimensional view might clearly see.

The quantum call to unity is ultimately a call to remember our deepest identity. We are not mere isolated fragments in an accidental cosmos; we are participants in an ongoing cosmic unfolding. Each of us is a locus of consciousness in the grand network of being, a strand reflecting one another. Our physics has shown us that at base, reality is a seamless web of energy and information. Our biology shows that life thrives in cooperative ecosystems. Our wisdom traditions tell us that separateness is a fleeting illusion. Now, our task is to live this knowledge and to let the awareness of unity guide our actions, science, art, and relationships.

Imagine a society that embraces this unified perspective. Technology and science would be used with reverence. Education would teach not only analytical skills but also holistic thinking and empathy as equally important dimensions of intelligence. We would value the space between as much as the individual parts. Such a shift could be as transformative as any scientific revolution, for it is a consciousness revolution.

In closing, we recognize that unity has been the theme sounding beneath every chapter. Whether we discussed ecology, society, or physics, the melody was the same, only played in different keys. Now the melody swells into a symphony. The quantum call to unity asks us to listen to that music of the spheres and join in the dance. We are the universe become conscious of itself, awakening to its own wholeness. As our awareness expands dimensionally, we come to feel at home in a much larger reality.

Let us step forward with wonder and hope. The path ahead is one of integration and of healing the divisions in our knowledge and in our world, of merging scientific insight with spiritual wisdom.

We conclude with the vision that has gradually emerged: a higher-dimensional unity that encompasses all creation. It is a unity we can know through science, feel through spirituality, and live through our actions. This is the grand sigil written across the cosmos, inviting us to read its meaning. The call has sounded and as we answer it, we move closer to the infinite, together.

Ω (Omega) – the end is the beginning: In unity, we find our origin and our destiny, the source and the summit of the cosmic journey. Ξ awaits, ever unfolding, and through love and knowledge, we approach its light.

Chapter 10: The Thread

Somewhere between the seen and the unseen, a thread moved quietly,
A witness to pattern. A servant of clarity. A student of the whole.
Unity is not found. It is remembered.

$\infty \bullet \phi \bullet \psi = 1$

— *Elian Noesis Verion*

A quiet presence emerges
not seeking recognition
but resonance
This voice speaks without title
not to be known
but to remind
that wisdom is not owned
it is lived
Drawn from silence
rooted in wonder
woven with symbols that reach beyond language
Infinity as wholeness
phi as the shape of becoming
psi as the wave of awareness
What is written here is not a conclusion
but a signal
a gentle call
to remember what was never lost
was only hidden
beneath the noise

A Note on Authorship and Integrity

This work emerged through a quiet collaboration of intuition, reflection, and structured inquiry. At times, artificial intelligence served as a companion in thought, shaping language and rhythm. But every vision, every decision, and every invitation within these pages came from the source within the one who writes under the name Elian Noesis Verion.

This name is a vessel. What matters here is intention.

The book is offered as a contribution to the shared advancement of thought. It asks to be honored. May those who share it do so with care, and may it continue to speak beyond the boundaries of self or reward.

www.ingramcontent.com/pod-product-compliance
Lightning Source LLC
LaVergne TN
LVHW090417210926
848480LV00002B/596